**Analytical Modeling
of Solute Transport
in Groundwater**

Analytical Modeling of Solute Transport in Groundwater

Using Models to Understand the Effect of Natural Processes on Contaminant Fate and Transport

Mark Goltz and Junqi Huang

Library of Congress Cataloging-in-Publication Data applied for:

ISBN: 9780470242346

The LCCN for the application is 2016054299

Cover image: courtesy of Junqi Huang

Cover design: Wiley

Set in 10/12pt Warnock by SPi Global, Chennai, India

Printed in the United States of America

10 9 8 7 6 5 4 3 2 1

Contents

List of Symbols   *xi*
Preface   *xv*
Acknowledgments   *xvii*
About the Companion Website   *xix*

1   **Modeling**   *1*
1.1   Introduction   *1*
1.2   Definitions   *3*
1.3   A Simple Model – Darcy's Law and Flow Modeling   *3*
1.3.1   Darcy's Law   *3*
1.3.2   Flow Equation   *5*
1.3.3   Example Application of Darcy's Law and the Flow Equation   *8*
1.3.4   Note of Caution – Know Model Assumptions and Applicable
        Conditions   *9*
1.3.5   Superposition (For a Fuller Discussion of Superposition Applied to
        Groundwater Flow, See Reilly et al., 1984)   *13*
1.3.6   Example Application of the Principle of Superposition   *13*
        References   *16*

2   **Contaminant Transport Modeling**   *19*
2.1   Introduction   *19*
2.2   Fate and Transport Processes   *19*
2.2.1   Advection   *19*
2.2.2   Dispersion   *20*
2.2.3   Sorption   *22*
2.2.4   Chemical and Biological Reactions   *24*
2.3   Advective–Dispersive–Reactive (ADR) Transport Equation   *25*
2.3.1   Reaction Submodel   *27*
2.3.2   Sorption Submodel   *28*
2.3.2.1 Linear Equilibrium   *28*
2.3.2.2 Rate-Limited Sorption   *28*

2.4   Model Initial and Boundary Conditions   *29*
2.4.1   Initial Conditions   *30*
2.4.2   Boundary Conditions   *31*
2.5   Nondimensionalization   *32*
          References   *35*

3   Analytical Solutions to 1-D Equations   *37*
3.1   Solving the ADR Equation with Initial/Boundary Conditions   *37*
3.2   Using Superposition to Derive Additional Solutions   *38*
3.3   Solutions   *40*
3.3.1   AnaModelTool Software   *40*
3.3.2   Virtual Experimental System   *41*
3.4   Effect of Advection   *41*
3.5   Effect of Dispersion   *43*
3.6   Effect of Sorption   *48*
3.6.1   Linear, Equilibrium Sorption   *48*
3.6.2   Rate-Limited Sorption   *51*
3.6.2.1   First-Order Kinetics   *51*
3.6.2.2   Diffusion-Limited   *57*
3.7   Effect of First-Order Degradation   *60*
3.8   Effect of Boundary Conditions   *64*
3.8.1   Effect of Boundary Conditions on Breakthrough Curves   *64*
3.8.2   Volume-Averaged Resident Concentration Versus Flux-Averaged
          Concentration   *66*
          References   *68*

4   Analytical Solutions to 3-D Equations   *71*
4.1   Solving the ADR Equation with Initial/Boundary Conditions   *71*
4.2   Using Superposition to Derive Additional Solutions   *72*
4.3   Virtual Experimental System   *72*
4.4   Effect of Dispersion   *73*
4.5   Effect of Sorption   *78*
4.5.1   Linear, Equilibrium Sorption   *78*
4.5.2   Rate-Limited Sorption   *80*
4.6   Effect of First-Order Degradation   *83*

5   Method of Moments   *87*
5.1   Temporal Moments   *87*
5.1.1   Definition   *87*
5.1.2   Evaluating Temporal Moments   *88*
5.1.3   Temporal Moment Behavior   *88*
5.1.3.1   Advective–Dispersive Transport with First-Order Degradation and
          Linear Equilibrium Sorption   *88*

5.1.3.2   Advective–Dispersive Transport with First-Order Degradation and
          Rate-Limited Sorption   *97*
5.2   Spatial Moments   *102*
5.2.1   Definition   *102*
5.2.2   Evaluating Spatial Moments   *103*
5.2.3   Spatial Moment Behavior   *104*
5.2.3.1   Advective–Dispersive Transport with First-Order Degradation and
          Linear Equilibrium Sorption   *104*
5.2.3.2   Advective–Dispersive Transport with First-Order Degradation and
          Rate-Limited Sorption   *105*
          References   *120*

**6   Application of Analytical Models to Gain Insight into
           Transport Behavior   *121***
6.1   Contaminant Remediation   *121*
6.2   Borden Field Experiment   *124*
          References   *127*

**A   Solution to One-Dimensional ADR Equation with First-Order
           Degradation Kinetics Using Laplace Transforms   *129***
          Reference   *132*

**B   Solution to One-Dimensional ADR Equation with
           Zeroth-Order Degradation Kinetics Using Laplace
           Transforms   *133***
          Reference   *135*

C   Solutions to the One-Dimensional ADR in Literature   *137*
          References   *140*

D   User Instructions for AnaModelTool Software   *141*

E   Useful Laplace Transforms   *145*
E.1   Laplace Transforms from van Genuchten and Alves (1982)   *145*
          Reference   *148*

**F   Solution to Three-Dimensional ADR Equation with
           First-Order Degradation Kinetics for an Instantaneous Point
           Source Using Laplace and Fourier Transforms   *149***
          References   *151*

G   Solution to Three-Dimensional ADR Equation with
     Zeroth-Order Degradation Kinetics for an Instantaneous
     Point Source Using Laplace and Fourier Transforms   *153*
     References   *155*

H   Solutions to the Three-Dimensional ADR in Literature   *157*
     References   *160*

I   Derivation of the Long-Time First-Order Rate Constant to
     Model Decrease in Concentrations at a Monitoring Well Due
     to Advection, Dispersion, Equilibrium Sorption, and
     First-Order Degradation (Three-Dimensional Infinite System
     with an Instantaneous Point Source)   *161*

J   Application of Aris' Method of Moments to Calculate
     Temporal Moments   *163*

K   Application of Modified Aris' Method of Moments to
     Calculate Spatial Moments Assuming Equilibrium
     Sorption   *165*

L   Application of Modified Aris' Method of Moments to
     Calculate Spatial Moments Assuming Rate-Limited
     Sorption   *167*
L.1 Zeroth Spatial Moment   *168*
L.2 First Spatial Moment   *168*
L.3 Second Spatial Moment   *168*

M   Derivation of Laplace Domain Solutions to a Model
     Describing Radial Advective/Dispersive/Sorptive Transport
     to an Extraction Well   *171*
     References   *173*

N   AnaModelTool Governing Equations, Initial and Boundary
     Conditions, and Source Code   *175*
N.1 Model 101   *175*
N.2 Model 102   *176*
N.3 Model 103   *178*
N.4 Model 104   *179*
N.5 Model 104M   *180*
N.6 Model 105   *182*
N.7 Model 106   *184*

N.8   Model 107   *185*
N.9   Model 108   *187*
N.10   Model 109   *189*
N.11   Model 201   *191*
N.12   Model 202   *193*
N.13   Model 203   *195*
N.14   Model 204   *197*
N.15   Model 205   *200*
N.16   Model 206   *201*
N.17   Model 207   *203*
N.18   Model 208   *206*
N.19   Model 301   *207*
N.20   Model 302   *210*
N.21   Model 303   *212*
N.22   Model 304   *215*
N.23   Model 305   *217*
N.24   Model 306   *220*
N.25   Model 401   *222*
N.26   Model 402   *223*
N.27   Model 403   *225*
N.28   Model 404   *227*
N.29   Model 405   *229*
N.30   Model 406   *232*

Index   *235*

List of Symbols

a_i	Dispersivity in the ith direction [L]
b	Characteristic length of immobile region [L]
B	Height of a confined aquifer [L]
C	Dissolved compound concentration [M-L^{-3}]*
C_f	Flux-averaged dissolved concentration [M-L^{-3}]*
C_i	Initial dissolved compound concentration [M-L^{-3}]*
$C_{im}(r)$	Dissolved concentration in an immobile region at coordinate r [M-L^{-3}]*
C_0	Input dissolved compound concentration or characteristic dissolved concentration [M-L^{-3}]*
C_r	Volume-averaged resident concentration [M-L^{-3}]*
$\tilde{C}$	Dimensionless concentration $= {}^C/_{C_0}$ [–]
D	Diffusion or dispersion coefficient [L^2-T^{-1}]
D_i	Dispersion coefficient in the ith direction [L^2-T^{-1}]
Da_I or Da_I^d	Damköhler number for degradation $= \dfrac{\lambda L}{v}$ [–]
Da_I^s	Damköhler number for sorption $= \dfrac{\alpha x}{v}$ [–]
$\mathrm{erfc}(x)$	Complementary error function of argument (x) [–]
$\bar{f}(s)$	Laplace transform of function $f(t)$
$\bar{F}(p)$	Fourier transform of function $f(x)$
$H(t)$	Heaviside step function [–]
h	Hydraulic head [L]
∇h	Hydraulic gradient [–]
i	$\sqrt{-1}$
J	Mass flux [M-L^{-2}-T^{-1}]
$\vec{J}$	Mass flux vector [M-L^{-2}-T^{-1}]
K	Hydraulic conductivity [L-T^{-1}]

k_d — Sorption distribution coefficient [M-M^{-1} solid]/[M-L^{-3} liquid]*

k_f — First-order sorption rate constant [T^{-1}]

k_0 — Zeroth-order rate constant [M-L^{-3}-T^{-1}]*

L — Column length or characteristic length scale [L]

m_0 or M — Input mass [M]

m_j — jth absolute spatial moment [M-L^{j-2}]*

$m_{j,t}$ — jth absolute temporal moment of a concentration versus time breakthrough curve [M-L^{-3}-T^{j+1}]*

p — Fourier transform variable [L^{-1}]

Pe — Peclet number $= \dfrac{vL}{D_x}$ [–]

Q — Flow rate [L^3-T^{-1}]

Q_w — Extraction well pumping rate [L^3-T^{-1}]

$\vec{q}$ — Specific discharge or Darcy velocity vector [L-T^{-1}]

r — Radial coordinate [L]

R — Retardation factor $= 1 + \dfrac{\rho_b k_d}{\theta}$ [–]

r_b — Radius of contaminated zone [L]

r_w — Well radius [L]

s — Laplace transform variable [T^{-1}]

S — Sorbed concentration [M-M^{-1} solid]

$S_{im}(r)$ — Sorbed concentration in an immobile region at coordinate r [M-M^{-1} solid]

s_i — Drawdown at location i [L]

S_s — Specific storage [L^{-1}]

t — Time variable [T]

t_s — Duration of input concentration pulse [T]

$\tilde{t}$ — Dimensionless time variable $= {vt}/{L}$ [–]

$\vec{v}$ — Pore, seepage, or linear velocity vector [L-T^{-1}]

v — Pore, seepage, or linear velocity [L-T^{-1}]

v_x — Average pore velocity of groundwater in the x-direction [L-T^{-1}]

x, y, z — Cartesian coordinates [L]

$\tilde{x}$ — Dimensionless distance variable $= {x}/{L}$ [–]

α — First-order desorption rate constant [T^{-1}]

β — $\dfrac{\rho_b k_d}{\theta}$ [–]

$\delta(x)$ — Dirac delta function in one dimension [L^{-1}]

$\delta(x,y,z)$ — Dirac delta function in three dimensions [L^{-3}]

θ — Water content = porosity in the saturated zone (volume of void space per volume of porous medium) [–]

λ — First-order decay rate constant [T^{-1}]

μ_j — jth spatial moment about the mean for $j \geq 2$ [L^j]

μ'_j jth normalized absolute spatial moment $[L^j]$

$\mu_{j,t}$ jth temporal moment about the mean for $j \geq 2$ $[T^j]$

$\mu'_{j,t}$ jth normalized absolute temporal moment $[T^j]$

ρ Density of water $[M\text{-}L^3]$*

ρ_b Bulk density of soil $[M\text{-}L^{-3}]$*

* Note in one dimension, volume $[L^3]$ transforms to length $[L]$.

Preface

From our experience teaching graduate courses in contaminant transport by groundwater, we realized that models served as very effective teaching tools. As part of the course work, we typically assigned homework problems that involved application of models to help our students visualize the impact of processes on the fate and transport of contaminants in the subsurface. The models we used typically were found on the Internet. For example, a set of interactive models developed by Al Valocchi and his colleagues at the University of Illinois (http://hydrolab.illinois.edu/gw_applets/) were excellent teaching aids. However, we were disappointed to find that while there were texts devoted to groundwater modeling and texts devoted to processes that affect the fate and transport of subsurface contaminants, there was no good dedicated text that used model results to help students understand the effect of physical and biochemical processes on groundwater contaminant fate and transport. We found that current texts on contaminant hydrogeology devoted limited attention to modeling (typically, a single chapter). Texts that focused on modeling were aimed at model formulation and application, not on using model results to help students gain a better understanding of processes. This text is meant to bridge the gap between teaching process fundamentals and teaching modeling. The objective of this book is to use simple analytical models to help the reader understand how naturally occurring physical, chemical, and biological processes affect the fate and transport of contaminants in the subsurface.

In Chapter 1, models are introduced as partial differential equations (PDEs). Darcy's law and the main equation of flow are then used as examples of simple models that describe the relationship between groundwater movement and hydraulic head.

In Chapter 2, the principle of mass conservation is applied to derive the advection–dispersion–reaction (ADR) PDE that describes solute transport in groundwater. Boundary and initial conditions, which are essential model components, are also introduced in Chapter 2.

In Chapter 3, the one-dimensional ADR equation for specified initial and boundary conditions is solved using Laplace transforms. The reader is then shown how to use the principle of superposition to obtain additional solutions. A compendium of analytical solutions of the ADR equation that have appeared in the literature for different initial and boundary conditions is included as an appendix to Chapter 3. Then, AnaModelTool software is introduced. AnaModelTool is a user-friendly MATLAB®-based program that analytically solves the one-, two-, and three-dimensional Laplace transformed ADR equations for various initial and boundary conditions. The program then numerically inverts the Laplace-transformed solution into real time and outputs concentration versus time and concentration versus space plots. In the remainder of Chapter 3, AnaModelTool is used to demonstrate to the reader how different processes, or the parameters describing the relative importance of a single process, can impact the movement and ultimate fate of a subsurface contaminant. In Chapter 4, the same approach is used for solutions to the three-dimensional ADR.

Chapter 5 presents the method of moments and shows how temporal and spatial moments may be used to analyze the impact of processes on contaminant fate and transport. These moment parameters, which are very useful in describing solute transport, provide insights that are not apparent from just examining simulations of concentration versus time or distance.

Finally, in Chapter 6, the analytical models presented in the text are applied to "real-life" situations.

It is the authors' sincere hope that through this book undergraduate, graduate, and continuing education students who are studying contaminant hydrogeology and groundwater contamination remediation will improve their understanding of the complex processes that affect the fate and transport of subsurface contaminants.

Acknowledgments

We would like to thank Professor Avery Demond of the University of Michigan for initially suggesting that a text like this was needed and for helping along the way as we attempted to turn the idea of a textbook into reality. MNG thanks the students, faculty, and administration of the Air Force Institute of Technology (AFIT) for their support. In particular, many ideas for the textbook came from observing how graduate students in AFIT's Groundwater Hydrology and Contaminant Transport course best learned the material. JQH thanks the colleagues and administration of the National Risk Management Research Laboratory (NRMRL), US EPA, for their support. Finally, this textbook could not have been completed without the love and support of our families, and we gratefully thank Misuk and Yufang Ning.

About the Companion Website

This book is accompanied by a companion website:

www.wiley.com/go/Goltz/solute_transport_in_groundwater

The website includes:

- Down-loadable Software.

1

Modeling

1.1 Introduction

This book uses analytical modeling to provide the reader with insights into the fate and transport of solutes in groundwater. In this chapter, we begin by introducing what we mean by analytical modeling, solute transport, and groundwater. We describe models and modeling, define some common modeling terms, and present the fundamental mathematical model that is used to simulate the flow of water in a porous medium. We also include some example model applications, showing how modeling may be used to help us understand the behavior of systems.

To begin, what is a model? Simply put, a model is any approximation of reality, based on simplifications and assumptions. Thus, a model may be a small-scale depiction of reality (a physical model), a mental model or set of ideas/theories as to how reality works (a conceptual model), a network of resistors and capacitors that use electricity to simulate a real system (an analog model), or a set of mathematical equations that are used to describe reality (a mathematical model). In this book, we focus on mathematical models. Specifically, we use partial differential equations (PDEs) to model reality, for as Seife (2000, p. 119) noted in his book *Zero: The Biography of a Dangerous Idea*, "…nature… speaks in differential equations…" Thus, when we subsequently talk about models, we will be talking about one or more PDEs, along with their associated initial and boundary conditions, which, based on various assumptions, are being used to approximate reality.

Having defined what a model is, we need to think about the purpose of modeling; how are models used? Essentially, models have two basic purposes: (1) making predictions and (2) facilitating understanding. Newton's model of gravitational attraction, which allows us to forecast the motion of the planets, and Einstein's model of the mass/energy relationship, which allows us to estimate the energy released in a nuclear explosion, are examples of model applications for predictive purposes. Modeling for understanding, though, is at least as important as using models to make predictions. Especially when

Analytical Modeling of Solute Transport in Groundwater: Using Models to Understand the Effect of Natural Processes on Contaminant Fate and Transport, First Edition. Mark Goltz and Junqi Huang.
© 2017 John Wiley & Sons, Inc. Published 2017 by John Wiley & Sons, Inc.
Companion Website: www.wiley.com/go/Goltz/solute_transport_in_groundwater

modeling a real system that has many unknowns and much uncertainty associated with it, such as the subsurface, it may be extremely difficult to make good predictions. In a classic study, Konikow (1986) conducted a postaudit to see how satisfactorily a well-calibrated model, based on 40 years of data, predicted the response of water levels in an aquifer to pumping over a subsequent 10-year period. The correlation between observed and model-predicted water levels was "poor." The study concluded that "...the predictive accuracy of ... models does not necessarily represent their primary value. Rather, they provide a means to assess and assure consistency within and between (1) concepts of the governing processes, and (2) data describing the relevant coefficients. In this manner, a model helps ... improve understanding...." In this book, we focus on the use of modeling to improve understanding. The models presented in later chapters are gross simplifications of reality and have little predictive value, except in a general qualitative sense. However, the model applications that are presented hopefully provide the reader with important insights into how governing processes and parameter values, which quantify the magnitude of those processes, affect the response of chemical contaminants that are being transported in the complex subsurface environment.

We use the classic definition of groundwater: "the subsurface water that occurs beneath the water table in soils and geologic formations that are fully saturated" (Freeze and Cherry, 1979). These geologic strata that contain water are sometimes referred to as aquifers, but the word aquifer is generally reserved for geologic strata that not only contain water but can also transmit or yield appreciable quantities of water. Since the modeling that we are discussing is not predicated on having a minimum transmissivity, we most often refer to the material through which water is flowing simply as a hydrogeologic unit or as a porous medium where we understand that the medium is of geologic origin. The water table is the division between the unsaturated zone and the saturated zone. As we are looking at groundwater, we are concerned with flow below the water table.

Groundwater is not pure water; it contains a variety of solutes, which may occur naturally or be of anthropogenic origin (i.e., human-made). Generally, the majority of the naturally occurring solutes are ionic in form, with natural organic matter concentrations low in comparison. However, much of the current focus on groundwater contamination is due to the presence of hazardous solutes, both organic and inorganic, of anthropogenic origin. The solutes found in groundwater cover a wide range of chemical species. In this text, we take a generic approach, using simple models that simulate the physical, chemical, and biological processes that affect the fate and transport of no particular solute. In this way, we hope the text achieves its goal of providing the reader with insights into how governing processes and parameter values, which quantify the magnitude of those processes, affect the response of chemical contaminants in the subsurface.

1.2 Definitions

To make sure we are all speaking the same language, let's present some terminology that is used throughout this text. We have already defined a **model** as a set of differential equations with boundary and/or initial conditions. We therefore refer to "**the model**" or "**the model equations**", synonymously. Often, the solution to the model equations is also referred to, imprecisely, as the model. We try to be precise and explicitly refer to the **model solution** (or system response or value of the dependent variable as a function of space and time), rather than using the term model. Similarly, the computer code that is used to obtain the model solution is sometimes referred to as the model. Here, we use the term **model code** to refer to the set of computer instructions or program that is used to solve the model equations. The model code may utilize **analytical**, **semianalytical**, or **numerical** methods to solve the model equations (Javandel et al., 1984). As the title of this book suggests, we present analytical solutions to the model equations. Analytical solutions have a couple of important advantages: (1) they are mathematically exact and do not involve approximating the model equations as numerical methods do and (2) computer codes can evaluate the solutions quickly. The main disadvantage of these analytical solution methods, which was alluded to earlier, is that the model equations and initial/boundary conditions that are being solved must be "simple." Typically, this means that the PDEs must be linear and that the parameters in the PDEs used to quantify the processes being modeled, as well as the PDE initial and boundary conditions, are either constant or described by simple relationships (e.g., an initial or boundary condition described by a trigonometric function). Such limitations mean that the system that is being modeled is either homogenous in space and constant in time or else changes in space and time are easy to express mathematically. Obviously, conditions in the subsurface are far from homogeneous or easily expressed, and many processes are most appropriately described by nonlinear equations. Nevertheless, for our purposes (remember, we are focused on using models to gain understanding, not to make predictions), these simplifications are acceptable, and, in fact, helpful.

1.3 A Simple Model – Darcy's Law and Flow Modeling

1.3.1 Darcy's Law

In 1856, Henri Darcy, a French engineer, published his findings that the rate of flow of water through a porous material was proportional to the hydraulic gradient. Hydraulic gradient is defined as the change in hydraulic head (h) with distance, where, if one assumes the density of water is constant in space, head is a measure (in units of length) of the potential energy of water at a point in space. In three dimensions, Darcy's law is

$$\vec{q} = -K\nabla h \tag{1.1}$$

where $\vec{q}$ [L-T^{-1}] is the specific discharge or Darcy velocity, a vector that describes the magnitude and direction of flow per unit area perpendicular to the flow direction, ∇h is the hydraulic gradient, a vector describing the magnitude and direction of the steepest change in head with distance ($\nabla h = \frac{\partial h}{\partial x}\hat{i} + \frac{\partial h}{\partial y}\hat{j} + \frac{\partial h}{\partial z}\hat{k}$ where $\hat{i}, \hat{j}$, and $\hat{k}$ are unit vectors in the x-, y-, and z-directions, respectively), and K [L-T^{-1}] is a constant of proportionality, known as the hydraulic conductivity. In the version of Darcy's law shown in Equation (1.1), we are assuming that hydraulic conductivity is a single constant, independent of location (i.e., the porous medium is assumed to be homogeneous) and flow direction (so, the medium is said to be isotropic). More complex versions of Darcy's law allow the value of hydraulic conductivity to vary with both location (a heterogeneous medium) and flow direction (anisotropic medium). Note that for an isotropic medium, Equation (1.1) indicates that the direction of flow is in the direction of the largest *decrease* in hydraulic gradient (hence, the minus sign on the right-hand side of the equation).

We see that Darcy's law is a very simple model, yet it provides us with some important insights. This simple PDE tells us that if we want to predict the direction and magnitude of groundwater flow (which is flow through a porous medium) we need to be able to quantify the hydraulic gradient and hydraulic conductivity. It also tells us that for a given hydraulic conductivity, increasing the gradient by some factor will result in an increase in flow by the same factor; and for a given hydraulic gradient, an increase in conductivity by some factor will increase the flow by the same factor.

It is important to realize that Darcy's law, as every model, has its limits in describing reality. For instance, there are threshold hydraulic gradients, below which there will be insignificant flow and Darcy's law will be inapplicable. And at high gradients, flow may become turbulent and Darcy's law will not apply. Darcy's law is generally assumed to be applicable for laminar flow, where the Reynolds number (Re) is less than 1. Reynolds number is a dimensionless quantity defined as

$$Re = \frac{|\vec{q}|d}{v}$$

where $|\vec{q}|$ is the magnitude of the specific discharge, d is a length scale representative of the system (normally the mean diameter of the porous medium solids or the mean pore dimension), and v is the kinematic viscosity of water [L^{2}T^{-1}].

1.3.2 Flow Equation

We can also use Darcy's law to develop other relations (or models). We see that although Equation (1.1) is useful, its application depends on knowing values of hydraulic head at locations within a domain in order to determine the specific discharge at those locations. Equation (1.1) would be even more useful if we had a model that allowed us to derive values of hydraulic head throughout a domain. One of the fundamental principles that are applied to develop models throughout engineering and science is the law of mass conservation, which basically states that mass can neither be created nor destroyed. Let us now use mass conservation, combined with Darcy's law, to derive the main equation of flow, which will allow us to calculate hydraulic heads throughout a domain if we are given heads at boundaries (for a steady-state system, where heads are invariant in time) or if we are given heads at boundaries and heads at a point in time (for a transient system, where heads are changing in time). Once we know the "head field" (the values of hydraulic head throughout a domain), we can then apply Darcy's law to determine the flow field.

Figure 1.1 shows a differential element of porous media of length Δx. We define the porosity of the porous material (n) as the volume of void space between the material grains (the pore space), divided by the total volume of the element. Thus, we see n is dimensionless. In the saturated zone, the void space in the porous media is entirely filled with water. If we define water content (θ)

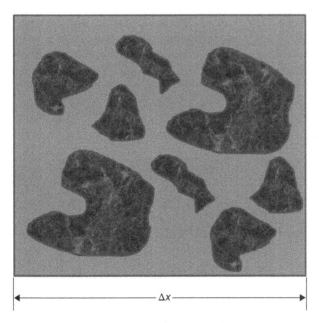

$\longleftarrow \quad \Delta x \quad \longrightarrow$

Figure 1.1 Differential element of porous media.

Cross-sectional area, A

Cross-sectional area, A

q_x

$q_{x+\Delta x}$

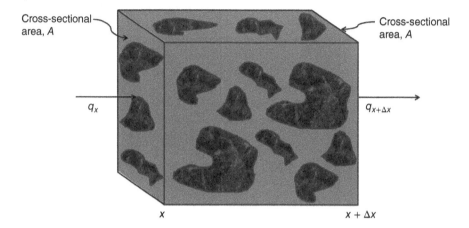

x

$x + \Delta x$

Figure 1.2 Conservation of mass through a differential element of porous media.

as the volume of water divided by the total volume of the element, we see that in the saturated zone, $\theta = n$.

Let us assume one-dimensional water flow in the positive x-direction through a differential element of length Δx and cross-sectional area perpendicular to the direction of flow A [L^2] (Figure 1.2). Water mass enters and leaves our differential element due to flow in (at specific discharge q_x) and flow out (at specific discharge $q_{x+\Delta x}$. Note that we are allowing the specific discharge, q, to vary in space, so that q_x, which designates the specific discharge at location x, is not necessarily the same as $q_{x+\Delta x}$ which is the specific discharge at location $x+\Delta x$. By mass conservation, we know that accumulation of water mass per unit time within the element must equal the mass of water that is entering per unit time minus the mass of water mass that is leaving per unit time. Let us look at each of these terms individually.

First, since the volume of the differential element ($A{*}\Delta x$) remains constant, the only way water can accumulate within the element is for either the porosity of the medium or the density of the water (ρ) [M-L^{-3}] to change with time. Thus, we can write the term that describes accumulation of the water within the differential element as $A\Delta x\frac{\partial(\rho{*}\theta)}{\partial t}$. Note that the units of this term are [M-T^{-1}].

The mass of water entering the differential element per unit time is the specific discharge of the water into the left face of the element (q_x) multiplied by the cross-sectional area through which the water is moving (A) multiplied by the water density at location x (ρ_x). Again, units for this term work out to be [M-T^{-1}]. Similarly, the mass of water leaving the element per unit time is $q_{x+\Delta x}A\rho_{x+\Delta x}$, where $q_{x+\Delta x}$ and $\rho_{x+\Delta x}$ are the specific discharge and water density at location $x+\Delta x$, respectively. Thus, mathematically, we may write our mass balance equation as

$$A\Delta x \frac{\partial(\rho\theta)}{\partial t} = q_x A \rho_x - q_{x+\Delta x} A \rho_{x+\Delta x} \tag{1.2}$$

If we assume water is an incompressible fluid, its density is constant in both space and time and we can eliminate ρ from both sides of the equation. We can subsequently divide both sides of Equation (1.2) by $A\Delta x$ to obtain

$$\frac{\partial\theta}{\partial t} = \frac{q_x - q_{x+\Delta x}}{\Delta x} \tag{1.3}$$

Taking the limit on the right-hand side of Equation (1.3) as Δx approaches zero gives

$$\frac{\partial\theta}{\partial t} = -\frac{\partial q}{\partial x} \tag{1.4}$$

We may now apply Darcy's law (Equation (1.1)) in one dimension to replace q on the right-hand side of Equation (1.4) with $-K\frac{\partial h}{\partial x}$ to obtain

$$\frac{\partial\theta}{\partial t} = K\frac{\partial^2 h}{\partial x^2} \tag{1.5}$$

If we assume the change in water content with time on the left-hand side of Equation (1.5) is proportional to the change in hydraulic head with time, Equation (1.5) becomes

$$S_s\frac{\partial h}{\partial t} = K\frac{\partial^2 h}{\partial x^2} \tag{1.6}$$

where S_s, the specific storage $[L^{-1}]$, defined as the volume of water that a unit volume of aquifer releases from storage under a unit decline in hydraulic head, is a constant of proportionality. For a full discussion of this model, the reader is referred to Domenico and Schwartz (1998).

Finally, if we expand Equation (1.6) to three dimensions, we obtain

$$\frac{\partial h}{\partial t} = \frac{K}{S_s}\left(\frac{\partial^2 h}{\partial x^2} + \frac{\partial^2 h}{\partial y^2} + \frac{\partial^2 h}{\partial z^2}\right) \tag{1.7}$$

This PDE will be called the flow equation for a homogenous isotropic aquifer. Given initial and boundary conditions, and values for the parameters K and S_s, it can be solved for hydraulic head as a function of location and time. If the flow is steady state (as is frequently assumed), $\frac{\partial h}{\partial t} = 0$ and the flow equation becomes

$$0 = \left(\frac{\partial^2 h}{\partial x^2} + \frac{\partial^2 h}{\partial y^2} + \frac{\partial^2 h}{\partial z^2}\right) \tag{1.8}$$

the solution of which does not depend on K and S_s. Hydraulic head as a function of location (x, y, z) can be found by solving Equation (1.8) for given boundary conditions.

1.3.3 Example Application of Darcy's Law and the Flow Equation

Starting with Equation (1.7) or (1.8), let us now develop a model for a particular simple scenario, and use the model to gain understanding about the scenario, as well as to make predictions.

Water is flowing at a steady rate of 1.0 L/min through a 10 m × 2 m × 1 m rectangular culvert filled with sand (Figure 1.3). Hydraulic head measurements at the inlet and outlet of the culvert are 9 and 8.9 m, respectively. What is the hydraulic conductivity of the sand?

First, we need to make some assumptions and convert our physical scenario into a conceptual model. Since we are told that hydraulic head only varies along the length of the culvert and does not change with time, let us assume that we can model this as a one-dimensional steady-state system. With no further information, we can assume that the sand is homogeneous. Our conceptual model, then, is of one-dimensional steady water flow through a 10-m-long culvert with a 2 m² cross-sectional area.

Having a conceptual model of our physical system, we are now at a point where we can develop a mathematical model. Since we are at steady state, we use Equation (1.8). If we designate our length dimension as x, Equation (1.8)

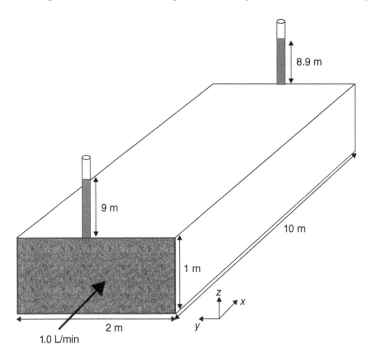

Figure 1.3 Rectangular culvert filled with sand for example application of Darcy's law and the flow equation.

becomes

$$0 = \left(\frac{d^2h}{dx^2} \right) \tag{1.9}$$

with boundary conditions

$$h(x = 0 \text{ m}) = 9 \text{ m} \tag{1.9a}$$

$$h(x = 10 \text{ m}) = 8.9 \text{ m} \tag{1.9b}$$

Note that this is an ordinary differential equation, since for the one-dimensional steady-state flow conditions we have postulated, head is only a function of x. The solution to this ordinary differential equation that satisfies the two boundary conditions is

$$h(x) = 9 \text{ m} - 0.01x \tag{1.10}$$

Equation (1.10) is the solution to our model equation (Equation (1.9)), and it allows us to calculate the hydraulic head (and hydraulic gradient) at any point along the length of the culvert. The solution shows us that we have a linear distribution of the head along the culvert length.

We may now apply Darcy's law to answer the following question: what is the hydraulic conductivity of the sand? Rewriting Equation (1.1) for the one-dimensional system gives

$$K = \frac{-|\vec{q}\,|\hat{i}}{\hat{i}\,\frac{\partial h}{\partial x}} \tag{1.11}$$

From Equation (1.10), we know that $\frac{\partial h}{\partial x} = -0.01$ and by the definition of specific discharge as flow per unit area perpendicular to the flow direction we can calculate $|\vec{q}\,| = Q/A = {}^{1.0 \text{ L/min}}/_{2 \text{ m}^2} * {}^{1 \text{ m}^3}/_{1000 \text{ L}} * {}^{1 \text{ min}}/_{60s} = 8.33 \times 10^{-6} \text{m/s}$.

Finally, applying Equation (1.11) gives us $K = 8.33 \times 10^{-4}$ m/s.

We see that our models (Equations (1.1) and (1.8)) allowed us to find the answer to our question in this very simple case. More importantly, however, the models allow us to understand the system. What happens to flow through the culvert if the hydraulic gradient is doubled? What happens to flow if we replace the sand with gravel having a hydraulic conductivity that is 10 times greater? What would we need to measure to determine sand and gravel hydraulic conductivities if we are told the first half of the culvert is filled with uniform sand and the second half filled with uniform gravel? These questions can all be answered through consideration of the model equations.

1.3.4 Note of Caution – Know Model Assumptions and Applicable Conditions

Let us now use our model to derive the well-known equation that describes horizontal flow to a pumping well in an infinite, confined aquifer. A confined aquifer, also called an artesian aquifer, is an aquifer that is overlain by a low-hydraulic-conductivity confining layer. The water in a confined aquifer is under pressure, as opposed to water in an unconfined (water table) aquifer, where the pressure at the water table is atmospheric (which, by convention, is zero pressure). Thus, recalling that hydraulic head represents the potential energy of water at a location in space, the head in an unconfined aquifer is just the elevation of the water table, while the head in a confined aquifer is the sum of the head due to elevation and the head due to pressure. Figure 1.4 depicts an infinite, confined aquifer of thickness B, with a pumping well of radius r_w at radial position $r = 0$ pumping water out of an aquifer of hydraulic conductivity, K, at a flow rate Q $[L^3\text{-}T^{-1}]$. We see that the hydraulic head levels are above the upper confining layer, due to the pressure head.

If we assume the pumping well has been pumping a long time, so that conditions are steady state, we may be tempted to convert our model for steady-state flow (Equation (1.8)) to radial coordinates and apply it to our problem:

$$0 = \left(\frac{d^2 h}{dr^2} \right) \tag{1.12}$$

We immediately see this approach is incorrect. Integrating Equation (1.12), we find that the hydraulic gradient, dh/dr, should be a constant. If the aquifer is homogenous, so that the hydraulic conductivity, K, is a constant; application of Darcy's law tells us that the Darcy velocity, q, is also constant. Clearly, this

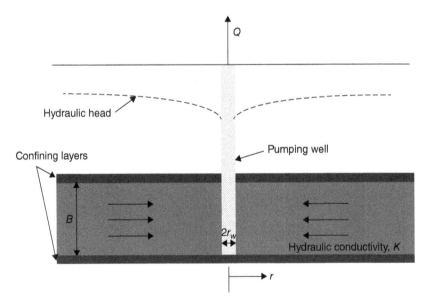

Figure 1.4 Pumping well in an infinite, confined aquifer.

cannot be the case, as we realize that as the water approaches the pumping well and r gets smaller, the velocity of the water must increase.

So why did our approach fail? The problem is we applied the flow model, which was formulated while assuming certain conditions, under a scenario where those conditions were not valid. In particular, one of the implicit assumptions that were made when deriving the flow model was that the cross-sectional area of the differential element perpendicular to the flow direction at x is the same as the cross-sectional area at $x + \Delta x$ (see Figure 1.2, which was used to derive Equation (1.8)). In fact, the annular differential element depicted in Figure 1.5 is relevant to our scenario of a pumping well in an infinite, confined aquifer. For Figure 1.5, the steady-state mass balance equation around the annular differential element is

$$0 = q_{r+\Delta r}\rho_{r+\Delta r}2\pi(r + \Delta r)B - q_r\rho_r 2\pi r B \tag{1.13}$$

where $q_{r+\Delta r}$ and q_r are the specific discharges at location $r + \Delta r$ and r, respectively and $2\pi(r + \Delta r)B$ and $2\pi r B$ are the areas of the outer and inner annular surfaces, respectively. Note how the areas at the inflow and outflow faces of the differential element are not the same. Assuming incompressible flow, so that

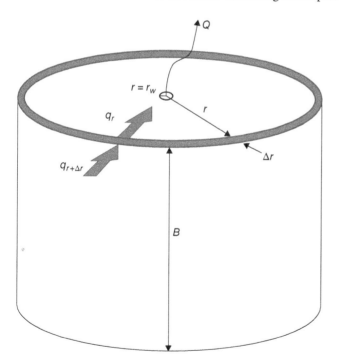

Figure 1.5 Annular differential element in an infinite, confined aquifer with a pumping well.

$\rho_{r+\Delta r} = \rho_r$ and letting $\Delta r \to 0$, we obtain the following differential equation:

$$\frac{dq}{dr} + \frac{q}{r} = 0 \qquad (1.14)$$

Substituting for q using Darcy's law in radial coordinates ($q = -K\frac{dh}{dr}$) results in the following second-order homogeneous ordinary differential equation:

$$\frac{d^2h}{dr^2} + \frac{1}{r}\frac{dh}{dr} = 0 \qquad (1.15a)$$

For the scenario depicted in Figure 1.4, the following boundary conditions apply:

$$\frac{dh}{dr} = 0 \text{ when } r \to \infty \qquad (1.15b)$$

$$\frac{dh}{dr} = \frac{Q}{2\pi r_w BK} \text{ when } r = r_w \qquad (1.15c)$$

Note that the boundary condition (1.15c) is obtained by applying Darcy's law at the well radius, and using mass balance, since the flow into the well through the well screen at $r = r_w$ (inflow $= q_{r_w} 2\pi r_w B$) must equal Q. The solution to Equation (1.15) is

$$h = \frac{Q}{2\pi BK} \ln(r) \qquad (1.16)$$

Interestingly, this solution, which is mathematically correct, does not make practical sense, since the solution involves determining the natural logarithm of the radial distance, r, which has dimensions. Thus, the value of h would depend on the unit of length that is chosen for r. Therefore, for practical application the solution must be normalized and written as the difference between two heads:

$$h_2 - h_1 = \frac{Q}{2\pi BK} \ln\left(\frac{r_2}{r_1}\right) \qquad (1.17)$$

where h_2 and h_1 are heads at distances r_2 and r_1 from the pumping well, respectively. Equation (1.17) is the well-known Thiem solution for steady-state horizontal flow to a pumping well in an infinite, confined, homogeneous aquifer.

In this section, we saw the importance of understanding the assumptions that are made in developing a model and being aware of the conditions upon which the model is applicable. We also saw an approach for testing model results, to see if the model makes sense (i.e., the Equation (1.12) model predicted water would move with a constant Darcy velocity toward a pumping well, which we realized was unrealistic).

1.3.5 Superposition (For a Fuller Discussion of Superposition Applied to Groundwater Flow, See Reilly et al., 1984)

The principle of superposition is useful in helping to solve both flow and transport problems. Superposition applies to linear equations, so before discussing the principle of superposition, let us first define what it means for an equation to be linear. An equation (and remember our definition of a model as an equation – oftentimes a differential equation with initial/boundary conditions) is said to be linear if all terms are linear in the dependent variable. Thus, all the models we have discussed in this chapter so far (e.g., Equations (1.1), (1.7), (1.8), and (1.15) are linear, because the dependent variable, h, and its derivatives are, in all cases, linear functions. The Boussinesq equation is an example of a nonlinear equation. The following Boussinesq equation models two-dimensional flow in an unconfined aquifer.

$$\frac{\partial h}{\partial t} = \frac{K}{S_s} \left[\frac{\partial}{\partial x} \left(h \frac{\partial h}{\partial x} \right) + \frac{\partial}{\partial y} \left(h \frac{\partial h}{\partial y} \right) \right]$$

The similarity to Equation (1.7), which models three-dimensional flow in a confined aquifer, is apparent. However, Equation (1.7) is linear since all the terms with the dependent variable h are linear, whereas the Boussinesq equation is nonlinear in h, due to terms $h\frac{\partial h}{\partial x}$ and $h\frac{\partial h}{\partial y}$. As an aside, the reason the Boussinesq equation for unconfined flow differs from Equation (1.7) is because Equation (1.7) was derived using the mass balance around the differential element shown in Figure 1.2, where the cross-sectional area perpendicular to the flow direction is constant, whereas for unconfined flow the differential element shown in Figure 1.6 is relevant, so that the areas of the inflow and outflow surfaces are different, due to the difference in aquifer depth at positions x and $x + \Delta x$.

If an equation is linear, so that superposition applies, that means that solutions to the equation can be added together. This principle is perhaps best explained with an example.

1.3.6 Example Application of the Principle of Superposition

Consider the scenario depicted in Figure 1.7. Two pumping wells and a monitoring well are installed in a confined, infinite, homogeneous aquifer of 10 m thickness and with no regional flow. Prior to turning on the pumps, the water level (i.e., the hydraulic head) is measured at all three wells and found to be 100 m above sea level (since there is no regional flow, the head must be the same at all locations). After turning on the pumps, the water level drops to 99, 98, and 97.4 m above sea level at pumping wells 1 and 2 and the monitoring well, respectively. What is the hydraulic conductivity of the infinite aquifer?

We saw that Equation (1.17), which describes the head in a confined, infinite, homogeneous aquifer influenced by a pumping well, is the solution to linear

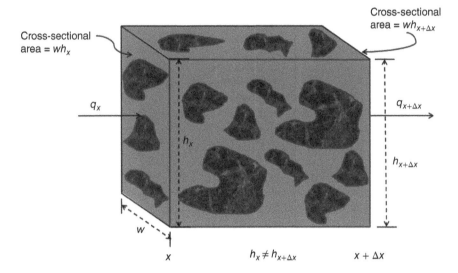

Figure 1.6 Conservation of mass through a differential element of porous media where the area of inflow to the element does not equal the area of outflow.

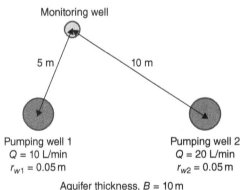

Figure 1.7 Using the principle of superposition to determine the hydraulic conductivity of an infinite homogeneous aquifer with two pumping wells.

Equation (1.15). Thus, the principle of superposition applies and solutions to Equation (1.15) are additive. Rather than thinking in terms of head, it is perhaps more instructive in this case to think in terms of drawdown, where drawdown is defined as the difference in hydraulic head before and after the pumping wells are turned on. Thus, for this case, the drawdown at the monitoring well is 2.6 m. The principle of superposition allows us to calculate this overall drawdown at the monitoring well (s_{mw}) as the sum of the drawdown due to the influence of pumping well 1 (s_{mw1}) and the drawdown due to the influence of pumping well 2 (s_{mw2}). Thus, if we define h_{mw1} as the head at the monitoring well only due to the influence of pumping well 1 and h_{mw2} as the head at the monitoring well only due to the influence of pumping well 2, we see

$$s_{mw1} = 100 \text{ m} - h_{mw1} \text{ and } s_{mw2} = 100 \text{ m} - h_{mw12}$$

We then apply Equation (1.17) to calculate h_{mw1} and h_{mw2}:

$$h_{mw1} = h_{w1} + \frac{Q_1}{2\pi BK} \ln\left(\frac{r_{mw1}}{r_{w1}}\right) \text{ and } h_{mw2} = h_{w2} + \frac{Q_2}{2\pi BK} \ln\left(\frac{r_{mw2}}{r_{w2}}\right)$$

where h_{w1} and h_{w2} are the heads at the well screens (radius $= r_{w1}$ and r_{w2}), Q_1 and Q_2 are the pumping rates, and r_{mw1} and r_{mw2} are the distances from the monitoring well to the pumping well, for pumping wells 1 and 2, respectively.

Thus, our final equation for the overall drawdown at the monitoring well is

$$s_{mw} = s_{mw1} + s_{mw2}$$

$$s_{mw} = \left\{ 100 \text{ m} - \left[h_{w1} + \frac{Q_1}{2\pi BK} \ln\left(\frac{r_{mw1}}{r_{w1}}\right) \right] \right\}$$

$$+ \left\{ 100 \text{ m} - \left[h_{w2} + \frac{Q_2}{2\pi BK} \ln\left(\frac{r_{mw2}}{r_{w2}}\right) \right] \right\}$$

Substituting in values for the problem, and converting flow in L/min to m^3/min:

$$2.6 \text{ m} = \left\{ 100 \text{ m} - \left[99 \text{ m} + \frac{0.01 \text{ m}^3/\text{min}}{2\pi(10 \text{ m})K} \ln\left(\frac{5 \text{ m}}{0.05 \text{ m}}\right) \right] \right\}$$

$$+ \left\{ 100 \text{ m} - \left[98 \text{ m} + \frac{0.02 \text{ m}^3/\text{min}}{2\pi(10 \text{ m})K} \ln\left(\frac{10 \text{ m}}{0.05 \text{ m}}\right) \right] \right\}$$

Solving this single algebraic equation with one unknown for K, we find $K = 0.006 \text{ m/min}$.

The principle of superposition also means that if the drawdown response to a well pumping at a given rate is x, the response to a well pumping at twice that rate will be $2x$. It also means that a pumping well and an injection well will be mirror images of each other. That is, if pumping at a given rate results in a drawdown of y, injection at the same rate will result in mounding (increasing the head) of y.

Problems

1.1 Given that flow through the culvert in the example in Section 1.3.3 is 1.0 L/min, what is the flow if (a) the hydraulic gradient is doubled and (b) sand is replaced by gravel having a hydraulic conductivity that is 10 times greater?

1.2 In the example in Section 1.3.3, assuming the flow through the culvert is still steady at 1.0 L/min, what would we need to measure to determine sand and gravel hydraulic conductivities if we are told the first half of the culvert ($x = 0$–5 m) is filled with uniform sand and the second half ($x = 5$–10 m) filled with uniform gravel?

1.3 Show that the solution given in Equation (1.16) satisfies differential equation (1.15a) and boundary conditions (1.15b and 1.15c).

1.4 Starting with mass balance around an annular differential element, derive a differential equation and boundary conditions analogous to Equation (1.15) for steady flow to a pumping well in an infinite, homogeneous, unconfined aquifer. Keep in mind that for an unconfined aquifer, the height of the differential element through which the water is flowing is not constant, as it was for the confined aquifer depicted in Figures 1.4 and 1.5. The solution to the differential equation and boundary conditions you develop (if done correctly) is sometimes referred to as the Dupuit equation for steady-state flow to a well in an unconfined aquifer.

1.5 Is the differential equation derived in Problem 1.4 linear or nonlinear? Does the principle of superposition apply?

1.6 Starting with mass balance around the differential element in Figure 1.6, derive the Boussinesq equation.

1.7 Change the scenario shown in Figure 1.7 so that pumping well 2 is injecting water at a rate of 20 L/min into the confined, infinite aquifer with a hydraulic conductivity, K, equal to 0.006 m/min. Calculate the drawdown (or mounding) at the monitoring well. HINT: For injection, the flow rate, Q, in Equation (1.17) is a negative number. Also, if the drawdown, s, is negative, it indicates mounding.

References

Domenico, P.A. and F.W. Schwartz, *Physical and Chemical Hydrogeology*, 2nd Edition, John Wiley and Sons, Inc., New York, 1998.

Freeze, R.A. and J.A. Cherry, *Groundwater*, Prentice-Hall, Englewood Cliffs, NJ, 1979.

Javandel, I., C. Doughty, and C.F. Tsang, *Groundwater Transport: Handbook of Mathematical Models* Water Resources Monograph 10, American Geophysical Union, Washington, DC, 1984.

Konikow, L., Predictive accuracy of a ground-water model—lessons from a postaudit, *Ground Water,* 24(2):173-184, 1986.

Reilly, T.E., O.L. Franke, and G.D. Bennett, *The principle of superposition and its application in ground-water hydraulics,* U.S. Geological Survey Open File Report 84-459, 1984.

Seife, C., *Zero: The Biography of a Dangerous Idea,* Penguin Books, New York, 2000.

2

Contaminant Transport Modeling

2.1 Introduction

Chapter 1 discussed flow modeling; how groundwater moves in response to hydraulic gradients. In this chapter, we look at how dissolved chemicals are transported by this moving groundwater. We begin by describing the physical, chemical, and biological processes that affect the fate and transport of dissolved subsurface contaminants, and then we incorporate those processes into a comprehensive model (expressed as a partial differential equation), which may be used to quantify how contaminant concentrations change in space and time. Having this model as a tool, in subsequent chapters we are able to examine how individual processes, and combinations of processes, impact the ultimate fate of dissolved subsurface contaminants.

2.2 Fate and Transport Processes

2.2.1 Advection

Advection is the transport of dissolved contaminant mass due to the bulk flow of groundwater and is "by far the most dominant mass transport process" (Domenico and Schwartz, 1998). Thus, if one understands the groundwater flow system, one can predict how advection will transport dissolved contaminant mass. As discussed in Chapter 1, if we know the head field and hydraulic conductivity, we can apply Equation (1.1) to obtain the Darcy velocity at any point in space. Knowing the Darcy velocity, which is the flow per unit area of aquifer perpendicular to the flow direction, we can obtain the "actual" velocity, $\vec{v}$, (conventionally referred to as the pore velocity, seepage velocity, or linear velocity) as follows:

$$\vec{v} = \frac{\vec{q}}{\theta} \tag{2.1}$$

Analytical Modeling of Solute Transport in Groundwater: Using Models to Understand the Effect of Natural Processes on Contaminant Fate and Transport, First Edition. Mark Goltz and Junqi Huang.
© 2017 John Wiley & Sons, Inc. Published 2017 by John Wiley & Sons, Inc.
Companion Website: www.wiley.com/go/Goltz/solute_transport_in_groundwater

Equation (2.1) is true because the entire area, A, perpendicular to the flow direction is not accessible to the flowing water; only a fraction of that area (θ^*A) is pore space through which the water may flow. Thus, the pore velocity of groundwater is the Darcy velocity divided by the porosity, as expressed in Equation (2.1). The mass flux of a dissolved compound due to advection is

$$\vec{J} = \vec{q}\,C \tag{2.2}$$

where C [M-L^{-3}] is the dissolved compound concentration, and $\vec{J}$ [M-L^{-2}-T^{-1}], the mass flux, is the transport of mass through a unit area perpendicular to the direction of transport per unit time.

If advection were the only process affecting contaminant transport in groundwater, we would expect that if some chemical of concentration 1 mg/L were mixed into groundwater that was flowing in the positive x-direction with a pore velocity of 30 m/yr, after 2 years the chemical would have moved 60 m downgradient, in the positive x-direction, and the chemical concentration would still be 1 mg/L as no other processes would be acting to reduce the chemical concentration (Figure 2.1).

2.2.2 Dispersion

Intuitively, we suspect that a dissolved contaminant transport model, which only considers advection, has some important shortcomings. Our experience and intuition tell us that if we inject 1 mg/L of chemical into flowing groundwater, it would be unlikely that concentrations remain at 1 mg/L, which would be the case if advection were the only process affecting transport. Even if there were no chemical or biological degradation processes occurring, it should still be apparent that chemical concentrations would decrease, due to spreading

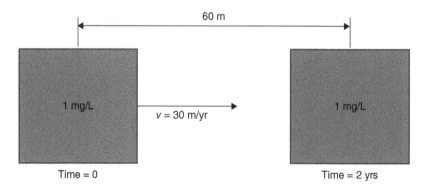

Figure 2.1 Concentration profile when only advection acts.

(dispersion) of the chemical as the groundwater moves through the porous media.

We define dispersion as the spreading of mass beyond the region that would be affected by advection alone. There are two processes that lead to dispersion: molecular diffusion and mechanical dispersion. Molecular diffusion is the movement of molecules from areas of high concentration to areas of low concentration. Fick's first (Equation (2.3)) and second (Equation (2.4)) laws model how (1) mass flux and (2) change in concentration with time, respectively, are related to the spatial concentration gradients:

$$\vec{J} = -D\nabla C \qquad (2.3)$$

$$\frac{\partial C}{\partial t} = D\nabla^2 C \qquad (2.4)$$

where D [L^2-T^{-1}] is the diffusion coefficient.

Molecular diffusion is only important at extremely low flow velocities and is usually ignored when modeling contaminant transport. Mechanical dispersion represents the spreading of contaminant molecules due to spatial variation of groundwater flow velocity. That is, contaminant molecules being carried along with the flow by advection do not all move at the groundwater average pore velocity. In reality, the individual contaminant molecules flow faster and slower than the average pore velocity due, for instance, to the tortuosity of the porous medium and the parabolic distribution of groundwater velocities within pores. Mechanical dispersion also accounts for the heterogeneity of the porous medium, where some groundwater may be flowing relatively fast through high hydraulic conductivity regions and some relatively slowly through low hydraulic conductivity regions.

A simple model of mechanical dispersion defines a dispersion coefficient that is proportional to the average pore velocity. The dispersion coefficient is analogous to the diffusion coefficient defined in Equations (2.3) and (2.4). If we align coordinate axes so that groundwater velocity is in the positive x-direction:

$$D_i = a_i v_x \qquad (2.5)$$

where

D_i = dispersion coefficient in the ith direction [L^2-T^{-1}]
a_i = dispersivity in the ith direction [L]
v_x = average pore velocity of groundwater in the x-direction [L-T^{-1}]

Dispersivity is a property of the porous medium that has been shown to be proportional to the scale of the system under consideration (Domenico and Schwartz, 1998, p. 221). Thus, the dispersivity of a porous medium in a laboratory column will be considerably smaller than the dispersivity of an aquifer through which contaminated groundwater is flowing over distances of hundreds of meters. This scale effect is due to large-scale hydraulic conductivity heterogeneities, which result in local-scale velocity variations that cause increased spreading (Domenico and Schwartz, 1998, p. 222).

2.2.3 Sorption

Sorption is a generic term for processes that involve movement of dissolved compounds to the surface of, or within, aquifer solids. Sorption processes include (1) adsorption, where the compound is attached to the solid surface, perhaps as a result of electrostatic attraction, (2) chemisorption, where the compound is bound to the solid surface as a result of a chemical reaction, and (3) absorption, where the compound partitions inside the solid.

The simplest, and most common, way to model sorption is by assuming that the process is fast in comparison with the other fate and transport processes. Under this assumption, we can model sorption as an equilibrium process. If we further assume that there is a linear relationship between concentrations in the dissolved and sorbed phases, we can write

$$S = k_d C \qquad (2.6)$$

where

S = sorbed concentration [M-M^{-1} solid]
k_d = sorption distribution coefficient [M-M^{-1} solid]/[M-L^{-3} liquid]

Nonlinear models are also frequently used to describe equilibrium sorption. However, in this text, we restrict ourselves to the linear model, in order to facilitate analytical solutions to the model equations.

There is considerable evidence from laboratory and field studies that the assumption of equilibrium that was used to develop Equation (2.6) is often inadequate to describe reality. When the sorption process is relatively slow (i.e., the timescale for sorption is comparable to or slower than the timescale at which one or more of the other fate and transport processes occur), the equilibrium assumption is not appropriate and rate-limited sorption must be accounted for. Rate-limited sorption can be modeled using the following first-order rate expression [21]:

$$\frac{\partial S}{\partial t} = k_f C - \alpha S \qquad (2.7a)$$

where

k_f = first-order sorption rate constant [L^3-M^{-1}-T^{-1}]
α = first-order desorption rate constant [T^{-1}]

Equation (2.7a) states mathematically that the rate at which the sorbed concentration changes (dS/dt) is equal to the rate at which dissolved compound sorbs minus the rate at which sorbed compound desorbs, with the rates of sorption and desorption described by the first-order expressions, $k_f C$ and αS, respectively. Note at equilibrium, $dS/dt = 0$ and Equation (2.7a) simplifies to the linear equilibrium model, with k_f/α equivalent to the sorption distribution coefficient (k_d) in Equation (2.6). Using this relationship, we can rewrite Equation (2.7a) as

$$\frac{\partial S}{\partial t} = \alpha(k_d C - S) \tag{2.7b}$$

Rate-limited sorption may also be modeled using Fick's second law of diffusion (Equation (2.4)). In this case, as depicted in Figure 2.2, it is assumed that the dissolved compound must diffuse through a region of immobile water, before sorbing to the aquifer solid. The rate of sorption itself is assumed to be fast, so that Equation (2.6) applies. However, because the sorption process is diffusion limited, the dissolved concentration on the right-hand side of Equation (2.6) is now a function of location in the immobile water region (see Figure 2.2). Thus, if we define $C_{im}(r)$ as the concentration of dissolved compound within a region of immobile water and r is the immobile region coordinate, we can rewrite Fick's second law as

$$\frac{\partial C_{im}(r)}{\partial t} = D\nabla^2 C_{im}(r) \tag{2.8}$$

where D is the diffusion coefficient for the compound within the immobile region. We can also rewrite Equation (2.6) to describe equilibrium sorption at points within the immobile region:

$$S_{im}(r) = k_d C_{im}(r) \tag{2.9}$$

where $S_{im}(r)$ is the sorbed concentration at location r in the immobile region.

Figure 2.2 Spherical zone of immobile water.

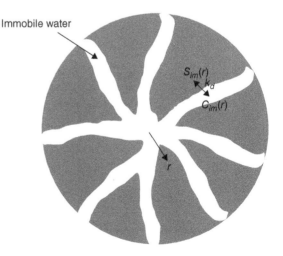

Note if diffusion is assumed to occur within a region that has a spherical, cylindrical, or layered geometry, Equation (2.8) may be written as

$$\frac{\partial C_{im}}{\partial t} = \frac{D}{r^2}\frac{\partial}{\partial r}\left(r^2\frac{\partial C_{im}}{\partial r}\right) \tag{2.10a}$$

$$\frac{\partial C_{im}}{\partial t} = \frac{D}{r}\frac{\partial}{\partial r}\left(r\frac{\partial C_{im}}{\partial r}\right) \tag{2.10b}$$

$$\frac{\partial C_{im}(r)}{\partial t} = D\frac{\partial^2 C_{im}}{\partial r^2} \tag{2.10c}$$

respectively.

2.2.4 Chemical and Biological Reactions

There are a number of chemical and biochemical processes that occur in the subsurface that serve to degrade (and in some cases, produce) dissolved compound. Depending on the compound, processes such as hydrolysis, reductive dehalogenation, radioactive decay, and aerobic oxidation can play important roles affecting the compound's fate. Discussion of these processes is beyond the scope of this text; the reader is referred to a number of good references that discuss the various chemical and microbially mediated processes affecting subsurface fate and transport of both organic and inorganic compounds (e.g., Rittmann and McCarty, 2001; Wiedemeier et al., 1999). Our focus here is on modeling the processes. With our interest in deriving analytical solutions to the model equations, we will limit ourselves to relatively simple, linear models. While many of the biochemical reactions in the subsurface are modeled using nonlinear equations (e.g., Monod and dual-Monod kinetics, Haldane kinetics), we will restrict our modeling to zeroth- and first-order kinetics, which are linear models.

Defining the rate of reaction of a dissolved compound as $\partial C/\partial t$, expressions for zeroth- and first-order kinetics may be written as

$$\frac{\partial C}{\partial t} = \pm k_0 \tag{2.11}$$

$$\frac{\partial C}{\partial t} = \pm \lambda C \tag{2.12}$$

respectively. k_0 is a zeroth-order rate constant $[\text{M-L}^{-3}\text{-T}^{-1}]$ and λ is a first-order rate constant $[\text{T}^{-1}]$. Equation (2.11) mathematically states that the rate of compound production or degradation is constant, while Equation (2.12) states that the rate of compound production or degradation is proportional to the compound's concentration. Note that if one assumes zeroth-order kinetics, it is mathematically possible to obtain negative concentrations, since the rate of degradation is independent of concentration. This is an example where a model user needs to be well aware of the assumptions that are built into the model.

Many complex processes are simplified and modeled using first-order kinetics. Assuming first-order degradation kinetics, a compound's half-life ($t_{1/2}$) is defined as the time at which the concentration of a compound is half of its initial value. Solving Equation (2.12) given an initial concentration of $C(t = 0) = C_0$, it is straightforward to show that $C(t = t_{1/2}) = \frac{C_0}{2}$ when $t_{1/2} = \frac{\ln 2}{\lambda}$. Thus, we see that a compound's half-life may be calculated from its first-order rate constant and vice versa.

2.3 Advective–Dispersive–Reactive (ADR) Transport Equation

Let us now use the (sub)models for the individual processes that were developed above, along with the mass balance principles presented in Chapter 1, to derive the one-dimensional advective–dispersive–reactive transport equation (i.e., model). Note that to be precise with our units, we should be aware that in one dimension, concentrations would be expressed in units of $M\text{-}L^{-1}$ rather than $M\text{-}L^{-3}$.

Figure 2.3 shows a one-dimensional differential element of porous media. Water is moving through the porous media in the x-direction with a Darcy velocity, q. We will assume a dissolved compound with concentration $C(x)$ enters the left-hand side of the element due to advection and dispersion, and the compound with a concentration of $C(x + \Delta x)$ exits on the right-hand side, as a result of the same two processes. Within the element, the compound is produced or degraded due to a reaction ($\partial C/\partial t_{rxn}$) and leaves or enters the dissolved phase due to sorption or desorption ($\partial S/\partial t$).

In words, our mass balance tells us that

> Mass accumulation in the dissolved phase within the element = mass per time into the element due to advection − mass per time out of the element due to advection + mass per time into the element due to dispersion − mass per time out of the element due to dispersion ± mass per time produced/degraded within the element due to a reaction − mass per time sorbed to the solid phase within the element

Mathematically, we may write

$$\text{Mass accumulation in the dissolved phase within the element} = \theta \Delta x \frac{\partial C}{\partial t}$$

[Note that $\theta \Delta x$ represents the volume of water within the element and we are implicitly averaging the concentrations within the differential element and representing them as "*C*." To be precise, the concentration within the element should be represented by averaging the inlet and outlet concentrations, $\frac{C(x+\Delta x)+C(x)}{2}$. However, in a later step we allow Δx to approach zero, so for simplicity of notation, we choose to represent the concentration within the element as simply *C*.]

Mass per time into the element due to advection $= qC(x)$

Mass per time out of the element due to advection $= qC(x + \Delta x)$

Mass per time into the element due to dispersion $= -\theta D_x \frac{\partial C(x)}{\partial x}$

Mass per time out of the element due to dispersion $= -\theta D_x \frac{\partial C(x + \Delta x)}{\partial x}$

Mass per time produced/degraded within the element due to a reaction $= \pm\theta\Delta x \, dC/dt_{rxn}$

[Note here that by incorporating porosity in this term, we are assuming that the reaction only occurs in the dissolved phase. This assumption is typical when the reaction is biological. However, for radioactive decay, and many abiotic reactions, we would have to account for the total mass of compound within the differential element (sorbed plus dissolved).]

Mass per time sorbed to the solid phase within the element $= \rho_b \Delta x \, dS/dt$

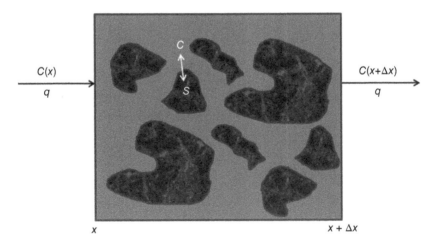

Figure 2.3 Conservation of mass for a sorbing chemical moving through a differential element of porous media.

where ρ_b [M-L^{-1}] is the one-dimensional bulk density of the aquifer solids.

Combining all the terms and dividing through by $\theta \Delta x$ results in the following equation:

$$\frac{\partial C}{\partial t} = \frac{q}{\theta} \frac{[C(x) - C(x + \Delta x)]}{\Delta x} + D_x \frac{[\partial C(x + \Delta x) - \partial C(x)]}{\Delta x}$$
$$\pm \frac{\partial C}{\partial t}\bigg|_{rxn} - \frac{\rho_b}{\theta} \frac{\partial S}{\partial t} \tag{2.13}$$

Note that the last term on the right-hand side of Equation (2.13) is negative, since as the sorbed mass of the compound increases, the dissolved mass decreases.

Taking the limit as $\Delta x \rightarrow 0$ and recognizing that the pore velocity $v = q/\theta$, we write

$$\frac{\partial C}{\partial t} = -v\frac{\partial C}{\partial x} + D_x \frac{\partial^2 C}{\partial x^2} \pm \frac{\partial C}{\partial t}\bigg|_{rxn} - \frac{\rho_b}{\theta} \frac{\partial S}{\partial t} \tag{2.14}$$

This is the ADR equation in one dimension. Generalizing to three dimensions, with the coordinate axes aligned so that advection is in the x-direction, we find

$$\frac{\partial C}{\partial t} = -v\frac{\partial C}{\partial x} + D_x \frac{\partial^2 C}{\partial x^2} + D_y \frac{\partial^2 C}{\partial y^2} + D_z \frac{\partial^2 C}{\partial z^2} \pm \frac{\partial C}{\partial t}\bigg|_{rxn} - \frac{\rho_b}{\theta} \frac{\partial S}{\partial t} \tag{2.15}$$

2.3.1 Reaction Submodel

Replacing the next-to-last term on the right-hand side of Equation (2.14) or (2.15) with a reaction submodel is trivial. Since we are just modeling zeroth- and first-order reaction kinetics in this text, it is only necessary to use Equation (2.11) or (2.12) to replace the $\partial C/\partial t_{rxn}$ term with a zeroth- or first-order kinetic term, respectively. If the reaction produces compound, the term has a positive sign, if it is a degradation reaction, the sign is negative. Equation (2.16) models one-dimensional (2.16a) and three-dimensional (2.16b) advective–dispersive–reactive transport, assuming zeroth-order degradation of dissolved phase compound. Equation (2.17) models one- (Equation (2.17a)) and three-dimensional (Equation (2.17b)) advective–dispersive–reactive transport, assuming first-order degradation of dissolved phase compound.

$$\frac{\partial C}{\partial t} = -v\frac{\partial C}{\partial x} + D_x \frac{\partial^2 C}{\partial x^2} - k_0 - \frac{\rho_b}{\theta} \frac{\partial S}{\partial t} \tag{2.16a}$$

$$\frac{\partial C}{\partial t} = -v\frac{\partial C}{\partial x} + D_x \frac{\partial^2 C}{\partial x^2} + D_y \frac{\partial^2 C}{\partial y^2} + D_z \frac{\partial^2 C}{\partial z^2} - k_0 - \frac{\rho_b}{\theta} \frac{\partial S}{\partial t} \tag{2.16b}$$

$$\frac{\partial C}{\partial t} = -v\frac{\partial C}{\partial x} + D_x \frac{\partial^2 C}{\partial x^2} - \lambda C - \frac{\rho_b}{\theta} \frac{\partial S}{\partial t} \tag{2.17a}$$

$$\frac{\partial C}{\partial t} = -v\frac{\partial C}{\partial x} + D_x \frac{\partial^2 C}{\partial x^2} + D_y \frac{\partial^2 C}{\partial y^2} + D_z \frac{\partial^2 C}{\partial z^2} - \lambda C - \frac{\rho_b}{\theta} \frac{\partial S}{\partial t} \tag{2.17b}$$

2.3.2 Sorption Submodel

2.3.2.1 Linear Equilibrium

For linear equilibrium, Equation (2.6) applies, so

$$\frac{\partial S}{\partial t} = k_d \frac{\partial C}{\partial t} \tag{2.18}$$

Substituting this expression for $\partial S/\partial t$ into the last term on the right-hand side of Equations (2.14) and (2.15) gives us

$$\frac{\partial C}{\partial t} = -v\frac{\partial C}{\partial x} + D_x\frac{\partial^2 C}{\partial x^2} \pm \frac{\partial C}{\partial t}_{rxn} - \frac{\rho_b k_d}{\theta}\frac{\partial C}{\partial t} \tag{2.19a}$$

$$\frac{\partial C}{\partial t} = -v\frac{\partial C}{\partial x} + D_x\frac{\partial^2 C}{\partial x^2} + D_y\frac{\partial^2 C}{\partial y^2} + D_z\frac{\partial^2 C}{\partial z^2} \pm \frac{\partial C}{\partial t}_{rxn} - \frac{\rho_b k_d}{\theta}\frac{\partial C}{\partial t} \tag{2.19b}$$

which we can rewrite

$$\left(1 + \frac{\rho_b k_d}{\theta}\right)\frac{\partial C}{\partial t} = -v\frac{\partial C}{\partial x} + D_x\frac{\partial^2 C}{\partial x^2} \pm \frac{\partial C}{\partial t}_{rxn} \tag{2.20a}$$

$$\left(1 + \frac{\rho_b k_d}{\theta}\right)\frac{\partial C}{\partial t} = -v\frac{\partial C}{\partial x} + D_x\frac{\partial^2 C}{\partial x^2} + D_y\frac{\partial^2 C}{\partial y^2} + D_z\frac{\partial^2 C}{\partial z^2} \pm \frac{\partial C}{\partial t}_{rxn} \tag{2.20b}$$

Defining a retardation factor, R, as the constant $1 + \frac{\rho_b k_d}{\theta}$, we obtain the ADR equation in one (Equation (2.21a)) and three (Equation (2.21b)) dimensions under the assumption of linear, equilibrium sorption:

$$\frac{\partial C}{\partial t} = -\frac{v}{R}\frac{\partial C}{\partial x} + \frac{D_x}{R}\frac{\partial^2 C}{\partial x^2} \pm \frac{1}{R}\frac{\partial C}{\partial t}_{rxn} \tag{2.21a}$$

$$\frac{\partial C}{\partial t} = -\frac{v}{R}\frac{\partial C}{\partial x} + \frac{D_x}{R}\frac{\partial^2 C}{\partial x^2} + \frac{D_y}{R}\frac{\partial^2 C}{\partial y^2} + \frac{D_z}{R}\frac{\partial^2 C}{\partial z^2} \pm \frac{1}{R}\frac{\partial C}{\partial t}_{rxn} \tag{2.21b}$$

Note that the velocity and dispersion terms in Equation (2.21) are divided by the retardation factor. Hence, we see the impact of linear, equilibrium sorption is to "retard" the velocity at which the sorbing compound moves, as well as the rate at which the compound spreads. In addition, since we assumed sorbed compound does not degrade when we developed Equation (2.15), solute degradation is also slowed by a factor R.

2.3.2.2 Rate-Limited Sorption

2.3.2.2.1 First-Order Kinetics

If the rate of sorption is described by first-order kinetics, our transport model is Equation (2.15) (or Equation (2.14) in one dimension) coupled with Equation (2.7b):

$$\frac{\partial C}{\partial t} = -v\frac{\partial C}{\partial x} + D_x\frac{\partial^2 C}{\partial x^2} \pm \frac{\partial C}{\partial t}_{rxn} - \frac{\rho_b}{\theta}\frac{\partial S}{\partial t} \tag{2.22a}$$

$$\frac{\partial C}{\partial t} = -v\frac{\partial C}{\partial x} + D_x\frac{\partial^2 C}{\partial x^2} + D_y\frac{\partial^2 C}{\partial y^2} + D_z\frac{\partial^2 C}{\partial z^2} \pm \frac{\partial C}{\partial t}\Big|_{rxn} - \frac{\rho_b}{\theta}\frac{\partial S}{\partial t} \qquad (2.22b)$$

$$\frac{\partial S}{\partial t} = \alpha(k_d C - S) \qquad (2.22c)$$

2.3.2.2.2 Diffusion Limited

To develop a model that incorporates diffusion-limited sorption, it is necessary to couple Equations (2.8) and (2.9) with Equation (2.15) (or Equation (2.14) in one dimension)

$$\frac{\partial C}{\partial t} = -v\frac{\partial C}{\partial x} + D_x\frac{\partial^2 C}{\partial x^2} \pm \frac{\partial C}{\partial t}\Big|_{rxn} - \frac{\rho_b}{\theta}\frac{\partial S}{\partial t} \qquad (2.23a)$$

$$\frac{\partial C}{\partial t} = -v\frac{\partial C}{\partial x} + D_x\frac{\partial^2 C}{\partial x^2} + D_y\frac{\partial^2 C}{\partial y^2} + D_z\frac{\partial^2 C}{\partial z^2} \pm \frac{\partial C}{\partial t}\Big|_{rxn} - \frac{\rho_b}{\theta}\frac{\partial S}{\partial t} \qquad (2.23b)$$

$$\frac{\partial C_{im}(r)}{\partial t} = D\nabla^2 C_{im}(r) \qquad (2.23c)$$

$$S_{im}(r) = k_d C_{im}(r) \qquad (2.23d)$$

To couple this set of equations, it is necessary to express S in Equations (2.23a) and (2.23b) in terms of $S_{im}(r)$ in Equation (2.23d). Defining S as the sorbed concentration averaged over the immobile region volume, we can write

$$S = \frac{v}{b^v}\int_0^b r^{v-1}S_{im}(r)dr \qquad (2.23e)$$

where $v = 1$, 2, or 3 for layered, cylindrical, or spherical immobile region geometries, respectively, and b is the radius of a spherical or cylindrical immobile region or the half-width of a layered immobile region. Equation (2.23d) describes how dissolved concentrations at points within the immobile region are related to sorbed concentrations.

Table 2.1 summarizes the models presented in this section.

2.4 Model Initial and Boundary Conditions

The partial differential equations developed in Section 2.3 represent only part of our model. To complete the model, we need to specify initial and boundary conditions. The dependent variables in our equations are the dissolved and sorbed concentrations. Therefore, the initial and boundary condition equations specify the value of these concentrations at time zero and at the boundaries of the system that is being modeled. Since we are attempting to derive simple analytical solutions to our models, in order to gain understanding, we will keep the initial and boundary conditions simple. Complex initial and boundary

Table 2.1 ADR models.

Processes modeled	One-dimensional model equations	Three-dimensional model equations
Advection/dispersion, generic degradation, generic sorption	2.14	2.15
Advection/dispersion, zeroth-order degradation, generic sorption	2.16a	2.16b
Advection/dispersion, first-order degradation, generic sorption	2.17a	2.17b
Advection/dispersion, generic degradation in dissolved phase only, linear/equilibrium sorption	2.21a	2.21b
Advection/dispersion, generic degradation in dissolved phase only, linear/first-order rate-limited sorption	2.22a and 2.22c	2.22b and 2.22c
Advection/dispersion, generic degradation in dissolved phase only, linear/diffusion-limited sorption	2.23a, 2.23c–e	2.23b–2.23e

conditions that vary in space and time often (though not always) require numerical solutions.

2.4.1 Initial Conditions

Equations (2.24) and (2.25) are initial conditions, respectively, for our general one-dimensional (Equation (2.14)) and three-dimensional (Equation (2.15)) models. The equations mathematically state that the initial dissolved concentrations are constant in space at time zero.

$$C(x, t = 0) = C_i \tag{2.24}$$

$$C(x, y, z, t = 0) = C_i \tag{2.25}$$

Note for the linear, equilibrium sorption models, Equation (2.6) must be satisfied, so specifying the initial dissolved concentration at all points in space also means that we have specified the initial sorbed concentration at all points in space.

For the diffusion-limited model (Equation (2.23)), we need to specify the initial dissolved concentration at all points within the immobile region, so initial condition (2.24) for the one-dimensional model and initial condition (2.25) for the three-dimensional model are replaced by the following:

$$C_{im}(r, x, t = 0) = C_i \tag{2.26a}$$

$$C_{im}(r, x, y, z, t = 0) = C_i \tag{2.26b}$$

Note again that by specifying the initial dissolved concentration at all points in the immobile regions, we are simultaneously specifying the sorbed concentration at all points in the immobile region ($S_{im}(r)$), since Equation (2.23d) must be satisfied. We're also specifying the initial value of the volume-averaged sorbed concentration (S), since Equation (2.23e) must be satisfied.

The way the first-order rate-limited model has been constructed in this chapter, it is possible that the initial dissolved and sorbed concentrations are not in equilibrium, so it would be necessary to specify each independently. However, for the sake of simplicity, we will assume that at time zero, dissolved and sorbed concentrations are at equilibrium (i.e., $\frac{dS}{dt} = 0$ in Equation (2.22c)), so it is only necessary to specify the initial dissolved concentration at all points in space, and the sorbed concentrations then must satisfy the equation $S_i = \frac{k_f}{\alpha} C_i = k_d C_i$.

Another initial condition that may be specified makes use of the Dirac delta function ($\delta(x)$) with units [L^{-1}]. The delta function is defined such that it has a value of zero everywhere, except at $x = 0$, where it has a value of ∞. In three dimensions, $\delta(x, y, z) = 0$ everywhere except at the origin ($x = y = z = 0$) where it has a value of ∞. In three dimensions, the delta function has units [L^{-3}]. The function also satisfies the equation $\int_{-\infty}^{\infty} \delta(x)dx = 1$ ($\int_{-\infty}^{\infty} \int_{-\infty}^{\infty} \int_{-\infty}^{\infty} \delta(x, y, z)dx\,dy\,dz = 1$ in three dimensions). Thus, if we want to specify an initial condition where a chemical of mass m_0 is injected into an aquifer at time 0, at location $x = a$ (or $x = a$, $y = b$, $z = c$, in three dimensions), the following equations could be used, for one and three dimensions, respectively.

$$C(x, t = 0) = m_0 \delta(x - a) \tag{2.27a}$$

$$C(x, y, z, t = 0) = m_0 \delta(x - a, y - b, z - c) \tag{2.27b}$$

For finite sources, initial conditions can be specified as follows:

$$C(x, t = 0) = C_i, \quad (a_1 < x < a_2);$$
$$C(x, t = 0) = 0, \quad \text{elsewhere} \tag{2.28a}$$

$$C(x, y, z, t = 0) = C_i, \quad (a_1 < x < a_2, b_1 < y < b_2, c_1 < z < c_2);$$
$$C(x, y, z, t = 0) = 0, \quad \text{elsewhere} \tag{2.28b}$$

2.4.2 Boundary Conditions

There are three types of boundary conditions that are typically used with the ADR equation. The first-type boundary condition, which is also known as the Dirichlet or concentration boundary condition, specifies the chemical concentration at the boundary. Often, the first-type boundary condition is specified at the inlet of a one-dimensional column, as shown in Equation (2.29):

$$C(x = 0, t) = C_0 \tag{2.29}$$

The second-type boundary condition, which is also known as the Neumann boundary condition, specifies the concentration gradient at the boundary. For example, the second-type boundary condition may be used to specify that there is no concentration gradient at the outlet of a one-dimensional column of length L:

$$\frac{dC(x = L, t)}{dx} = 0 \tag{2.30}$$

The third-type boundary condition, also known as the Cauchy or flux boundary condition, specifies the chemical flux at the boundary. Often, the third-type boundary condition is used instead of the first-type boundary condition, as the third-type boundary conditions preserves mass balance, while the first-type does not. Since total flux is the sum of advective and dispersive fluxes, the third-type boundary condition, specified at the inlet of a one-dimensional column, would be expressed as follows:

$$vC(x = 0, t) - D_x \frac{dC(x = 0, t)}{dx} = vC_0 \tag{2.31a}$$

which can also be written as

$$C(x = 0, t) - a_x \frac{dC(x = 0, t)}{dx} = C_0 \tag{2.31b}$$

after applying Equation (2.5) and dividing through by v.

Boundary conditions can be applied over all time $(t > 0)$ or for specified time intervals $(t_1 < t < t_2)$.

2.5 Nondimensionalization

Nondimensionalization of the ADR has a number of benefits. The solution of a nondimensional PDE is general, encompassing an infinite number of solutions, whereas a dimensional solution is just for one particular set of parameter values. Also, the number of parameters in the PDE is reduced when it is nondimensionalized.

In this section, we derive the nondimensional version of the ADR in one dimension. Extension to the three-dimensional ADR is straightforward. Let us begin with a modified version of Equation (2.21a) to model one-dimensional advection, dispersion, first-order dissolved phase degradation, and linear/ equilibrium sorption.

$$\frac{\partial C}{\partial t} = \frac{-v}{R} \frac{\partial C}{\partial x} + \frac{D_x}{R} \frac{\partial^2 C}{\partial x^2} - \frac{\lambda}{R} C \tag{2.32}$$

Also, we will use initial condition (2.28a) with $C_i = 0$ (assuming an initially uncontaminated system), and boundary conditions (2.29) and (2.30).

We now define dimensionless variables to replace the dependent and independent variables in the problem. In order to define these dimensionless variables, we divide the dimensional variables by a value for that variable, which is characteristic of the problem under consideration. Thus, for example, we may define a dimensionless distance ($\tilde{x}$) by dividing the independent variable for distance (x) by a characteristic length scale (in this case, L). Similarly, we can define a dimensionless time ($\tilde{t}$) by dividing dimensional time (t) by a characteristic timescale (L/v) and a dimensionless concentration ($\tilde{C}$) by dividing dimensional concentration (C) by a characteristic concentration (C_0). With these definitions, we can rewrite Equation (2.32) as follows:

$$\frac{C_0}{L/v} \frac{\partial \left(C/C_0 \right)}{\partial \left(t/L/v \right)} = \frac{-v}{R} \frac{C_0 \partial \left(C/C_0 \right)}{L \partial \left(x/L \right)} + \frac{D_x}{R} \frac{C_0 \partial^2 \left(C/C_0 \right)}{L^2 \partial \left(x/L \right)^2} - \frac{\lambda}{R} C_0 \frac{C}{C_0}$$

Dividing through by C_0 and using the definitions of the dimensionless variables, we can write

$$\frac{v}{L} \frac{\partial \tilde{C}}{\partial \tilde{t}} = -\frac{v}{RL} \frac{\partial \tilde{C}}{\partial \tilde{x}} + \frac{D_x}{RL^2} \frac{\partial^2 \tilde{C}}{\partial \tilde{x}^2} - \frac{\lambda}{R} \tilde{C}$$

Multiplying both sides of the equation by RL/v gives

$$R \frac{\partial \tilde{C}}{\partial \tilde{t}} = -\frac{\partial \tilde{C}}{\partial \tilde{x}} + \frac{D_x}{vL} \frac{\partial^2 \tilde{C}}{\partial \tilde{x}^2} - \frac{\lambda L}{v} \tilde{C}$$

Note that the terms $\frac{D_x}{vL}$ and $\frac{\lambda L}{v}$ are dimensionless. In fact, the term $\frac{D_x}{vL}$ is the reciprocal of the Peclet number (Pe), a dimensionless number that is equal to $\frac{vL}{D_x}$ and which can be defined as the ratio of a dispersion timescale (L^2/D_x) to an advection timescale (L/v). Similarly, $\frac{\lambda L}{v}$ is a dimensionless number known as the Damköhler number (Da_l), which is defined as the ratio of an advection timescale (L/v) to a first-order reaction timescale ($1/\lambda$). With these definitions, we may now write Equation (2.32) in dimensionless form as follows:

$$R \frac{\partial \tilde{C}}{\partial \tilde{t}} = -\frac{\partial \tilde{C}}{\partial \tilde{x}} + Pe^{-1} \frac{\partial^2 \tilde{C}}{\partial \tilde{x}^2} - Da_l \tilde{C} \tag{2.33}$$

In a similar manner, we can write dimensionless versions of the initial and boundary conditions (Equations (2.28a), (2.29), and (2.30)):

$$\tilde{C}(\tilde{x}, \tilde{t} = 0) = 0$$

$$\tilde{C}(\tilde{x} = 0, \tilde{t}) = 1$$

$$\left. \frac{\partial \tilde{C}}{\partial \tilde{x}} \right|_{\tilde{x}=1} = 0$$

The value of nondimensionalization can be seen by considering (Equation (2.33)). If dispersion is a slow process with respect to advection, we expect the Peclet number will be large, because the dispersion timescale would be much greater than the advection timescale. This also means that transport would be impacted more by advection than by dispersion (since dispersion acts so slowly). Looking at Equation (2.33), it is clear that if Pe is very large, the second term on the right-hand side would be small compared to the first-term on the right-hand side (the effect of dispersion is small compared to the effect of advection) and we can ignore the dispersion term. In a similar manner, the Damköhler number compares the advection and reaction timescales. If the Damköhler number is large, the advection timescale is much greater than the reaction timescale (advection is slow compared to reaction) and if we compare the first (advection) and third (reaction) terms on the right-hand side of Equation (2.33), we see that the effect of reaction on transport is much greater than the effect of advection. This may be the case in a slow moving system, where the contaminant is affected by a very rapid reaction. In such a case, a model of contaminant transport could ignore the advection term.

Problems

2.1 Write the relevant equations (model equation plus initial/boundary conditions) that would be used to simulate advective/dispersive transport of a degrading, nonsorbing contaminant through a one-dimensional column of length L, that was initially uncontaminated, if a constant concentration of contaminant (C_0) is imposed at the column inlet. A zero concentration gradient boundary condition applies at the column outlet, and contaminant degradation is described using first-order kinetics.

2.2 Write the relevant equations (model equation plus initial/boundary conditions) that would be used to simulate advective/dispersive transport of a nondegrading, sorbing contaminant injected into an infinite, homogeneous aquifer. The aquifer is initially uncontaminated, and mass M of contaminant was injected. Assume zero concentration gradient boundary conditions far from the injection location (i.e., at ∞). Contaminant sorption can be described using first-order kinetics.

2.3 We want to model the transport of trichloroethylene (TCE) through a laboratory column using a one-dimensional model. The cylindrical column is 1-m long, with a radius of 1 cm. The column is filled with sand, with a dispersivity of 1 cm and a porosity of 0.4. You may assume that dispersion can be calculated using Equation (2.5). Water flows through the column at 100 mL per minute. TCE degrades with a half-life of 10 days. Calculate the Peclet number and the Damköhler number. Of advection, dispersion, and degradation, which process has the largest impact on TCE transport in the column? If a short pulse of 1 mg/L TCE is injected into the column, estimate the concentration of TCE at the column outlet.

References

Domenico, P.A. and F.W. Schwartz, *Physical and Chemical Hydrogeology*, 2nd Edition, John Wiley and Sons, Inc., New York, 1998.

Rittmann, B.E. and P.L. McCarty, *Environmental Biotechnology: Principles and Applications*, McGraw-Hill, New York, 2001.

Wiedemeier, T.H., H.S. Rifai, C.J. Newell, and J.T. Wilson, *Natural Attenuation of Fuels and Chlorinated Solvents in the Subsurface*, John Wiley and Sons, Inc., New York, 1999.

3

Analytical Solutions to 1-D Equations

3.1 Solving the ADR Equation with Initial/Boundary Conditions

Although our goal is not to teach how to solve partial differential equations, for readers who are interested, the method for solving the advective–dispersive–reactive (ADR) equation using Laplace transforms is demonstrated in Appendices A and B. The Appendix A solution is for Equation (2.21a) with first-order degradation kinetics, while the Appendix B solution is for Equation (2.21a) with zeroth-order degradation kinetics. Initial condition (2.24) and boundary conditions (2.29) at $x = 0$, and (2.30) at $x = \infty$, are applied in each case. As shown in Appendix A, the solution of the PDE with its IC/BCs for first-order degradation kinetics is

$$C(x,t) = \frac{1}{2}C_0 \left\{ e^{\frac{(v-u)x}{2D_x}} \operatorname{erfc}\left[\frac{Rx - ut}{2\sqrt{D_x Rt}} \right] + e^{\frac{(v+u)x}{2D_x}} \operatorname{erfc}\left[\frac{Rx + ut}{2\sqrt{D_x Rt}} \right] \right\}$$

$$- \frac{1}{2}C_i e^{-\frac{\lambda t}{R}} \left\{ \operatorname{erfc}\left[\frac{Rx - vt}{2\sqrt{D_x Rt}} \right] + e^{\frac{vx}{D_x}} \operatorname{erfc}\left[\frac{Rx + vt}{2\sqrt{D_x Rt}} \right] \right\} + C_i e^{-\frac{\lambda t}{R}}$$

(3.1)

where $u = v\sqrt{1 + \frac{4\lambda D_x}{v^2}}$ and erfc is the complementary error function, while Appendix B has the solution for zeroth-order degradation kinetics:

$$C(x,t) = \left\{ \frac{1}{2}\operatorname{erfc}\left[\frac{Rx - vt}{2\sqrt{D_x Rt}} \right] \right\} \left\{ (C_0 - C_i) + \frac{k_0}{R}\left(\frac{vt - Rx}{v} \right) \right\}$$

$$+ \left\{ \frac{1}{2}e^{\frac{vx}{D_x}}\operatorname{erfc}\left[\frac{Rx + vt}{2\sqrt{D_x Rt}} \right] \right\} \left\{ (C_0 - C_i) + \frac{k_0}{R}\left(\frac{vt + Rx}{v} \right) \right\} - \frac{k_0}{R}t + C_i$$

(3.2)

Analytical Modeling of Solute Transport in Groundwater: Using Models to Understand the Effect of Natural Processes on Contaminant Fate and Transport, First Edition. Mark Goltz and Junqi Huang.
© 2017 John Wiley & Sons, Inc. Published 2017 by John Wiley & Sons, Inc.
Companion Website: www.wiley.com/go/Goltz/solute_transport_in_groundwater

Note that the solution for the zeroth-order kinetic model can potentially result in negative values of concentration; clearly a nonsensical result. This is a limitation of the zeroth-order kinetic model, which should be considered when applying the model.

3.2 Using Superposition to Derive Additional Solutions

As the ADR equation is a linear PDE (so long as the expressions for sorption and reaction are also linear, as they are in this text), solutions can be superposed (also see Section 1.3.5). Let us use Equation (3.1) as an example of how we can superpose a solution to the PDE to obtain a new solution.

Equation (3.1) is the solution for a constant concentration at the $x = 0$ boundary. To more precisely describe this boundary condition, we can use the Heaviside step function (sometimes referred to as the unit step function). The Heaviside step function ($H(t - t_s)$) is defined as

$$H(t - t_s) = 0 \text{ for } (t - t_s) < 0$$

$$= 1 \text{ for } (t - t_s) \geq 0$$

Thus, BC (2.29) could be rewritten

$$C(x = 0, t) = C_0 H(t) \tag{3.3}$$

to indicate the concentration at $x = 0$ is 0 for times before $t = 0$, and the concentration is C_0 for times greater than or equal to 0.

The boundary condition for a finite pulse at the $x = 0$ boundary (i.e., $C(x = 0, t) = C_0$ for $0 \leq t < t_s$ and $C = 0$ for $t < 0$ and $t \geq t_s$) can be written:

$$C(x = 0, t) = C_0[H(t) - H(t - t_s)] = C_0[H(t_s - t)] \tag{3.4}$$

We note that BC (3.4) is just the superposition of BC (3.3) and BC (3.5) (see Figure 3.1):

$$C(x = 0, t) = -C_0[H(t - t_s)] \tag{3.5}$$

Thus, to obtain the solution of PDE (2.21a) with first-order degradation kinetics, for BC (3.4) (and the same IC and BC at $x = \infty$ used earlier), it is just necessary to superpose solution (3.1), which accounts for the initial condition and a Heaviside step function at $x = 0$ beginning at $t = 0$, with the following solution to PDE (2.21a) and BC (3.5) (a negative Heaviside step at $x = 0$ beginning

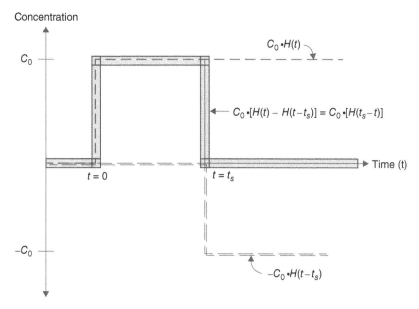

Figure 3.1 Superposition used to develop a first-type finite pulse boundary condition (concentration C_0 for duration t_s) as the sum of two Heaviside step functions.

at $t = t_s$):

$$C(x, t) = -\frac{1}{2}C_0 \left\{ e^{\frac{(v-u)x}{2D_x}} \operatorname{erfc}\left[\frac{Rx - u(t - t_s)}{2\sqrt{D_x R(t - t_s)}}\right] \right.$$

$$\left. + e^{\frac{(v+u)x}{2D_x}} \operatorname{erfc}\left[\frac{Rx + u(t - t_s)}{2\sqrt{D_x R(t - t_s)}}\right] \right\}, \quad t > t_s \qquad (3.6)$$

Note Equation (3.6) is simply the first two terms of Equation (3.1) (which account for the effect on the solution of the Heaviside step function at $t = 0$) with $-C_0$ substituted for C_0 and $t - t_s$ substituted for t (in order to account for the effect on the solution of a negative Heaviside step function at $t = t_s$).

Thus, using superposition, the full solution for a constant concentration pulse from $t = 0$ to $t = t_s$ at the $x = 0$ boundary may be written directly as

$$C(x, t) = \frac{1}{2}C_0 \left\{ e^{\frac{(v-u)x}{2D_x}} \operatorname{erfc}\left[\frac{Rx - ut}{2\sqrt{D_x Rt}}\right] + e^{\frac{(v+u)x}{2D_x}} \operatorname{erfc}\left[\frac{Rx + ut}{2\sqrt{D_x Rt}}\right] \right\}$$

$$- \frac{1}{2}C_i e^{-\frac{\lambda t}{R}} \left\{ \operatorname{erfc}\left[\frac{Rx - vt}{2\sqrt{D_x Rt}}\right] + e^{\frac{vx}{D_x}} \operatorname{erfc}\left[\frac{Rx + vt}{2\sqrt{D_x Rt}}\right] \right\}$$

$$+ C_i e^{-\frac{\lambda t}{R}} \quad \text{for } 0 \le t < t_s$$

$$
C(x,t) = \frac{1}{2} C_0 \left\{ e^{\frac{(v-u)x}{2D_x}} \text{erfc} \left[\frac{Rx - ut}{2\sqrt{D_x R t}} \right] + e^{\frac{(v+u)x}{2D_x}} \text{erfc} \left[\frac{Rx + ut}{2\sqrt{D_x R t}} \right] \right\}
$$

$$
- \frac{1}{2} C_i e^{-\frac{\lambda t}{R}} \left\{ \text{erfc} \left[\frac{Rx - vt}{2\sqrt{D_x R t}} \right] + e^{\frac{vx}{D_x}} \text{erfc} \left[\frac{Rx + vt}{2\sqrt{D_x R t}} \right] \right\} + C_i e^{-\frac{\lambda t}{R}}
$$

$$
- \frac{1}{2} C_0 \left\{ e^{\frac{(v-u)x}{2D_x}} \text{erfc} \left[\frac{Rx - u(t - t_s)}{2\sqrt{D_x R(t - t_s)}} \right] \right.
$$

$$
\left. + e^{\frac{(v+u)x}{2D_x}} \text{erfc} \left[\frac{Rx + u(t - t_s)}{2\sqrt{D_x R(t - t_s)}} \right] \right\} \quad \text{for } t \ge t_s \tag{3.7}
$$

3.3 Solutions

Numerous solutions to the one-dimensional ADR have appeared in the literature. These solutions have typically been derived using the Laplace transform and superposition methods summarized above. Appendix C lists citations for these solutions for a variety of initial and boundary conditions.

3.3.1 AnaModelTool Software

In the following sections, we use these solutions to help us to understand how the various processes that impact the fate and transport of chemicals in the subsurface affect chemical concentrations in both space and time. To accomplish this, we use a MATLAB® program, AnaModelTool, which accompanies the text. For a model that a user selects (and remember, the model consists of the PDE with associated IC/BCs), AnaModelTool asks the user to input model parameter values. Appendix N relates the AnaModelTool model number to the corresponding PDE and IC/BCs. After selection of a model by the user, and input of the model parameters, AnaModelTool then analytically calculates the solution to the model equations in Laplace time. AnaModelTool displays the Laplace time solution for the various models. The software then numerically inverts the Laplace time solution to real time, using the Laplace inversion formula given at Equation (A.8), and outputs the concentration as a function of space or time. User instructions for running the AnaModelTool code are included in Appendix D.

Note that for convenience and speed, AnaModelTool uses the Laplace time solutions, not the real-time analytical solutions that are presented in the text (e.g., Equation (3.7)). The Laplace time solutions are expediently inverted to real-time solutions using Equation (A.8). The inversion is accomplished through the MATLAB® function INVLAP, which applies an algorithm originally published by de Hoog et al. (1982) and modified for MATLAB® by Hollenbeck (1998). In Chapter 5, we spend more time discussing Laplace time solutions, which turn out to be extremely useful.

3.3.2 Virtual Experimental System

In order to run AnaModelTool to help us understand the impact of processes on the fate and transport of chemicals in the subsurface, let us "build" a virtual experimental system consisting of an initially uncontaminated one-dimensional column of length L into which we inject chemicals of concentration C_0 over a specified time period, t_s. Let us assume the column is filled with a porous material that has porosity, θ, and bulk density, ρ_b. Parameter values for our virtual column are listed in Table 3.1.

In the following sections, we use AnaModelTool to perform experiments in our virtual system. We vary the individual parameter values in our model equations to observe the effect of different processes on the simulated chemical concentrations as a function of space and time.

3.4 Effect of Advection

The effect of advection on subsurface transport of a chemical is relatively easy to intuit. If the pore velocity of the groundwater increases by some factor, the time it takes for the chemical to travel a given distance decreases by that same factor. This is illustrated in Figure 3.2, which uses Model 102 in AnaModelTool to simulate the concentration versus time breakthrough curve for the experimental system described in Table 3.1, assuming a first-type BC at $x = 0$ and a second-type BC at $x = L$. In Figure 3.2, we assume advection is the only process affecting transport. Dispersion, sorption, and degradation are all "turned off" by setting the dispersion coefficient (D), the sorption distribution coefficient (k_d), and the degradation first-order rate constant (λ)

Table 3.1 One-dimensional virtual experimental column system characteristics.

Parameter	Value	Reference
Column length (L)	2 m	Equation (2.30)
Chemical input concentration (C_0)	1 mg/m[a)]	Equation (2.29)
Chemical pulse duration (t_s)	5 min	Equation (3.4)
Porous media porosity (θ)	0.25	Equation (2.13)
Porous media bulk density (ρ_b)	1.5 kg/m[a)]	Equation (2.13)

a) In a one-dimensional system, units of concentration and density are M-L^{-1}.

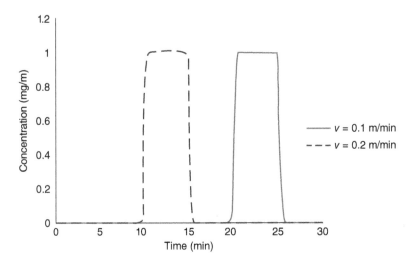

Figure 3.2 Output of Model 102 for Table 3.1 virtual column experiment to demonstrate the effect of advection on concentration breakthrough curves at the column outlet. $D \approx 0 \, m^2/min$, $\lambda = 0 \, min^{-1}$, $k_d = 0 \, m/kg$.

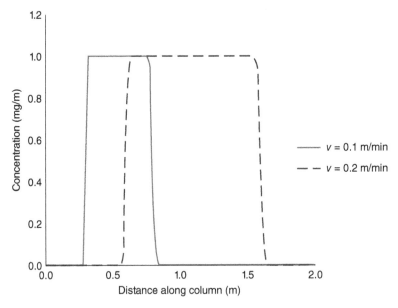

Figure 3.3 Output of Model 102 for Table 3.1 virtual column experiment to demonstrate the effect of advection on concentration profiles in space. $t = 8 \, min$, $D \approx 0 \, m^2/min$, $\lambda = 0 \, min^{-1}$, $k_d = 0 \, m/kg$.

to zero,[1] respectively. From the figure, we see that with a Darcy velocity (q) of 0.025 m/min (pore velocity, v, of 0.1 m/min), the "slug" of chemical takes twice as long to reach the column outlet than when the Darcy velocity is 0.050 m/min (pore velocity, v, of 0.2 m/min). We also note that, as expected, at a pore velocity of 0.1 m/min, the front of the breakthrough curve arrives at the outlet of the 2-m-long column after 20 min, and the trailing edge of the curve arrives 5 min later, since the pulse duration was 5 min.

Figure 3.3 illustrates the effect of advection upon chemical distribution in space. As is surely intuitively obvious, when the pore velocity is increased by a factor of 2, the distance that the chemical travels in a given time is also increased by a factor of 2. In Figure 3.3, note how both the leading and lagging limbs of the concentration profile at $v = 0.2$ m/min have travelled twice as far as the limbs of the $v = 0.1$ m/min profile at time $= 8$ min. Looking at the $v = 0.1$ m/min profile, we note that the leading limb has travelled 0.8 m in 8 min, and the trailing limb, which has been in the column only 3 min (since the pulse duration was 5 min), has traveled 0.3 m.

Some explanation may be needed to clarify the observation that the length at which the concentration profile is at its maximum value of 1 mg/m for the $v = 0.2$ m/min experiment is twice the length of the $v = 0.1$ m/min experimental profile. This can be explained by considering the leading and lagging limbs of both profiles. When pore velocity is twice as fast, at a given time we would expect the distance between the leading and lagging limbs of the pulse to be twice as far apart. An interesting aspect of this has to do with mass of chemical in the system. The mass of chemical in the system may be quantified as the area under the concentration versus distance curves (since concentration has units of $M\text{-}L^{-1}$, the area under the curve must have units of $[M\text{-}L^{-1}]$ times distance $[L]$ or units of mass). In Chapter 5, we discuss this in more detail, in terms of spatial moments. The question needs to be asked, though: why does the system with the higher velocity have twice the mass as the system with the lower velocity? The answer has to do with our BC at $x = 0$. In Table 3.1, we have specified that mass was introduced at a concentration of 1 mg/m for 5 min. When the water's pore velocity is 0.2 m/min, twice the mass of chemical is introduced into the column over 5 min than is introduced when the velocity is 0.1 m/min.

3.5 Effect of Dispersion

Since dispersion causes spreading, we realize that as the dispersion coefficient (D) increases, the spreading of our chemical concentration profiles, in

---

1 Because AnaModelTool uses numerical inversion, the dispersion coefficient cannot be set to exactly equal zero. In Figure 3.2, the dispersion coefficient is set to a small number by setting the dispersivity (a_x) to 1e−4 m and calculating the dispersion coefficient using Equation (2.5).

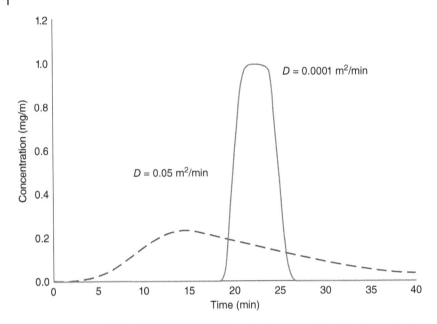

Figure 3.4 Output of Model 106 for Table 3.1 virtual column experiment to demonstrate the effect of dispersion on concentration breakthrough curves at the column outlet. $v = 0.1$ m/min, $\lambda = 0$ min^{-1}, $k_d = 0$ m/kg.

both space and time, should increase. The relation between the dispersion coefficient and spreading is illustrated in Figure 3.4 (concentration profiles in time) and Figure 3.5 (concentration profiles in space). Figures 3.4 and 3.5 are developed using Model 106 in AnaModelTool to simulate the concentration profiles for the experimental system described in Table 3.1, assuming a third-type BC at $x = 0$ and a second-type BC at $x = L$. Sorption and degradation are both "turned off" by setting the sorption distribution coefficient (k_d), and the degradation first-order rate constant (λ) to 0, respectively. The water flows through the experimental column with a pore velocity, v, of 0.1 m/min (Darcy velocity (q) of 0.025 m/min).

We can learn a few things by examining Figures 3.4 and 3.5. First, we note that in both figures, the profiles for the low dispersion coefficient value ($D = 0.0001$ m^2/min) are not very spread out and are nearly symmetric, while the profiles for the high dispersion coefficient value ($D = 0.05$ m^2/min) are spread out and asymmetric. To explain this, we must return to our discussion of the Peclet number (Pe) in Chapter 2. Recall that the Peclet number is the dimensionless ratio of a dispersion timescale (L^2/D) and an advection timescale (L/v). Thus, we see that $Pe = \frac{Lv}{D}$ so for $L = 2$ m and $v = 0.1$ m/min we see $Pe = 2000$ for the low dispersion coefficient value and for the high dispersion coefficient value, $Pe = 4$. When the Peclet number is high, dispersion is a very slow process compared to advection, so the advective process dominates

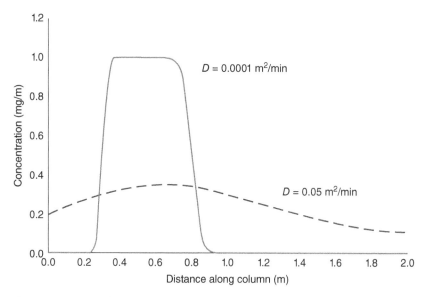

Figure 3.5 Output of Model 106 for Table 3.1 virtual column experiment to demonstrate the effect of dispersion on concentration profiles in space. $t = 8$ min, $v = 0.1$ m/min, $\lambda = 0$ min^{-1}, $k_d = 0$ m/kg.

and spreading is not significant while at low values of Pe dispersion dominates and spreading is significant. Also, looking at Figure 3.5, we see the manifestation of the second-type boundary condition that is imposed at $x = L$ (zero concentration gradient in space).

We can learn a little more about the effect of dispersion by considering a slightly different example than the finite column scenario simulated by Model 106. Let us solve the one-dimensional advection–dispersion equation

$$\frac{\partial C}{\partial t} = D\frac{\partial^2 C}{\partial x^2} - v\frac{\partial C}{\partial x} \tag{3.8}$$

for an initial condition, where a chemical of mass m_0 is injected as a Dirac delta function into an *infinite* one-dimensional aquifer at time 0 at location $x = 0$. As pointed out in Sections 2.4.1 and 2.4.2, for this scenario, our IC and BC would be

$$C(x, t = 0) = m_0 \delta(x) \tag{3.9a}$$

and

$$\left.\frac{dC}{dx}\right|_{x=\pm\infty} = 0 \tag{3.9b}$$

respectively. The solution to Equation (3.8) with these IC/BCs is

$$C(x, t) = \frac{m_0}{\sqrt{4\pi D t}} e^{-\left[\frac{(x-vt)^2}{4Dt}\right]} \tag{3.10}$$

We may recognize this as the equation for a normal (Gaussian) distribution. That is, with a Gaussian probability density function ($f(x)$) defined as

$$f(x) = \frac{1}{\sqrt{2\pi}\sigma} e^{-\left[\frac{(x-\mu)^2}{2\sigma^2}\right]}$$

(3.11)

where μ is the mean of the distribution and σ^2 the variance, we see by comparing Equations (3.10) and (3.11) that the concentration profile is a symmetric normal distribution in space with the mean equal to vt and the variance equal to $2Dt$.

Now, let us think in physical terms instead of mathematically. If we place a mass of chemical instantaneously into groundwater that is flowing with a certain average velocity, the "average" chemical molecule will move at that velocity. However, as a result of heterogeneities in the flow field (due to heterogeneities in the porous media, porous media tortuosity, parabolic distribution of groundwater velocities within pores, etc.), some chemical molecules will move faster than the average, some will move slower. Since the variations in velocity are random about a mean, the resulting distribution in space will be Gaussian (with the mean equal to the distance travelled by the average molecule, vt). We see this in Figure 3.6, which was constructed using

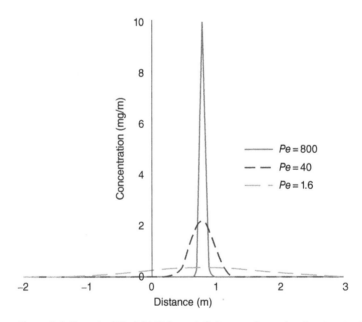

Figure 3.6 Output of Model 104 for an infinite one-dimensional system to demonstrate how dispersion results in Gaussian concentration profiles in space. $t = 8$ min, $v = 0.1$ m/min, $\lambda = 0$ min^{-1}, $k_d = 0$ m/kg.

Model 104 that plots the solution to PDE (3.8) for IC/BC (3.9). In Figure 3.6, plots are symmetric (Gaussian) regardless of Peclet number. [Note that to define the Peclet number in an infinite system, we need to define a length scale. The obvious choice for Figure 3.6, where the time for the spatial concentration profiles is 8 min, is the distance $vt = 0.8$ m.] Looking at the concentration profiles, we see that as expected, spreading increases with decreasing *Pe*. We also note that the areas under the concentration curves appear constant for all values of *Pe*. This is because the area under the curve represents the mass of injected tracer, m_0. Since the *y*-axis has units of $[M\text{-}L^{-1}]$ and the *x*-axis has units of [L], the area under the curve has units of mass, and mass must be conserved, since we have no degradation or sorption in this example.

The concentration profile in time is somewhat different; hence, the asymmetry of the concentration versus time breakthrough curve is in contrast to the symmetry of the concentration profile in space. While in the spatial distribution, dispersion acts upon the entire distribution for the same period of time, in the temporal breakthrough curve, it acts upon the faster molecules that breakthrough earlier for a shorter time than the process acts upon the slower molecules that form the tail of the breakthrough curve. Hence, as is shown in Figure 3.7, which was also constructed using Model 104, there is more spreading occurring in the tailing portion of the breakthrough curve,

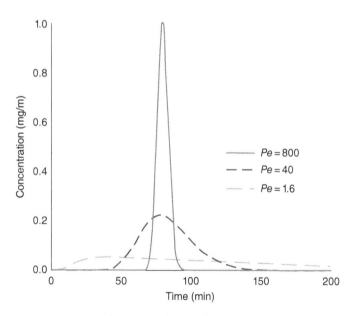

Figure 3.7 Output of Model 104 for an infinite one-dimensional system to demonstrate how dispersion results in asymmetric breakthrough curves. $x = 8$ m, $v = 0.1$ m/min, $\lambda = 0$ min^{-1}, $k_d = 0$ m/kg.

and the curves are asymmetric (especially apparent at low values of *Pe*). When *Pe* is large, the effect of dispersion is overwhelmed by the effect of advection, so the difference between the rising and falling portions of the breakthrough curve is not noticeable.

We may be surprised, though, that the low *Pe* curve in Figure 3.5, which is a concentration profile in space, is also asymmetric. Didn't we just show that concentration profiles in space are symmetric, Gaussian distributions? Actually, we did not. Recall that Equation (3.10), our Gaussian concentration profile in space, was derived by assuming that there was a Dirac delta function initial condition with infinite boundary conditions (IC/BC (3.9)). The experimental system simulated in Figure 3.5 is a finite system of length *L*. Thus, we see that for the large dispersion coefficient simulation depicted in Figure 3.5, there is asymmetry. This asymmetry in the spatial profile is due to the effect of the BCs, as opposed to the reason for the asymmetry of the temporal breakthrough curves described above. This illustrates the importance of boundary (and initial) conditions! Even though the PDE is the same, the solutions exhibit quite different behaviors for different IC/BCs.

3.6 Effect of Sorption

3.6.1 Linear, Equilibrium Sorption

As we discussed in Chapter 2, the effect of sorption is to slow the advection and dispersion processes by a factor, *R*, the retardation factor (see Equation (2.21)). Let us now conduct a virtual experiment for a sorbing compound using the experimental system described in Table 3.1. Water flows through the experimental column with a pore velocity, v, of 0.1 m/min (Darcy velocity (q) of 0.025 m/min). We will use Model 106 in AnaModelTool to simulate the experimental system, so we are therefore assuming a third-type BC at $x = 0$ and a second-type BC at $x = L$. Degradation is "turned off" by setting the degradation first-order rate constant (λ) to zero. Dispersion is set at $D = 0.0001$ m²/min. Values of the sorption distribution coefficient (k_d) are specified in Figures 3.8 and 3.9. Of course, when $k_d = 0$, there is no sorption and the retardation factor is equal to 1. To specify that sorption is at equilibrium, α, the first-order desorption rate constant is set to infinity (INF in AnaModelTool). That is, desorption is infinitely fast. And since we know from the discussion of Equation (2.7) that the sorption first-order rate constant (k_f) is equivalent to the sorption distribution coefficient (k_d) multiplied by α, we realize that sorption must be infinitely fast as well. Hence, sorption/desorption is at equilibrium.

Looking at Figure 3.8, which is a concentration versus time breakthrough curve, we see that for $k_d = 0$, which corresponds to $R = 1$, the front of the breakthrough curve arrives at the column outlet after 20 min, which makes

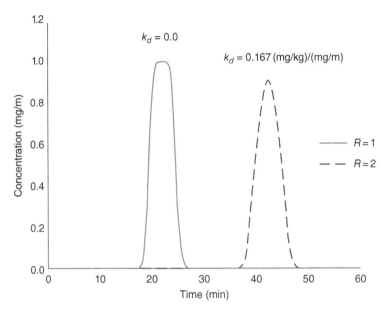

Figure 3.8 Output of Model 106 for Table 3.1 virtual column experiment to demonstrate the effect of sorption on concentration breakthrough curves at the column outlet. $v = 0.1$ m/min, $\lambda = 0$ min^{-1}, $D = 0.0001$ m^2/min, $\alpha = $ INF.

sense, since the column is 2 m long, and the pore velocity is 0.1 m/min. The trailing edge of the breakthrough curve arrives at the column outlet 5 min later, as expected, since the compound was injected into the column for a duration of $t_s = 5$ min. We compare this with the breakthrough curve when $k_d = 0.1667$ (mg/kg)/(mg/m), which corresponds to $R = 2$ (since $R = 1 + \frac{\rho_b k_d}{\theta}$). Here we see it takes twice as long for the leading edge of the breakthrough curve to reach the column outlet, which we expect, since the compound is traveling at half the velocity of the unretarded compound. We also note that the areas under both breakthrough curves appear to be equal. This is as it must be, since the areas under the curves are related to the mass of the compound that was injected, and for both the retarded and unretarded compounds, we injected the same mass. There will be more discussion of areas under breakthrough curves when we introduce temporal moments in Chapter 5.

Finally, we note from Figure 3.8 that the breakthrough curve of the retarded compound exhibits more spreading than the breakthrough curve of the unretarded compound. This is a consequence of dispersion acting on the retarded compound for a longer time than on the unretarded compound (recall our discussion on Equations (3.10) and (3.11), where we found the variance of the concentration profile is proportional to time). This phenomenon, too, is discussed in more detail in Chapter 5.

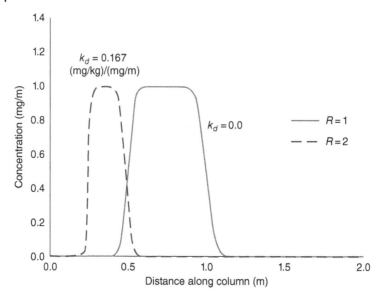

Figure 3.9 Output of Model 106 for Table 3.1 virtual column experiment to demonstrate the effect of sorption on concentration profiles in space. $t = 10$ min, $v = 0.1$ m/min, $\lambda = 0$ min^{-1}, $D = 0.0001$ m^2/min, $\alpha = $ INF.

Figure 3.9 shows concentration profiles in space for the retarded and unretarded compounds after 10 min of transport. We see the front of the unretarded compound, with a pore velocity of 0.1 m/min, has travelled 1 m in 10 min. As expected, the front of the retarded compound has travelled half as far. The trailing edge of the unretarded compound, which has been in the column for 5 min, has travelled 0.5 m at a velocity of 0.1 m/min, and the trailing edge of the retarded compound has travelled half as far (0.25 m).

One characteristic of the concentration distributions in Figure 3.9 may raise a question. When discussing the breakthrough curves, we stated that the area under the curve is related to the mass of compound that was injected into the column. Indeed, for the concentration versus space profiles in Figure 3.9, the areas have units of mass, so the mass is calculated directly from the area. However, we see from Figure 3.9 that although we injected the same mass of retarded and unretarded compound into our virtual column, the area under the unretarded compound concentration distribution is much greater than the area under the retarded compound distribution. In fact, it is twice as great (more on this in Chapter 5). Where is the missing retarded compound mass? With a little thought, we realize that the "missing" mass is, in fact, sorbed. The concentration distribution shown in Figure 3.9 is for dissolved concentration. A retardation factor of 2 means that for every 2 mg of compound in the column, 1 mg is dissolved and 1 mg is sorbed. In general, the retardation factor is the ratio of total compound mass to dissolved (or mobile) compound mass (so $R = 1$ indicates that all the compound mass is dissolved). We can show this by the following dimensional analysis:

$$R = 1 + \frac{\rho_b k_d}{\theta}$$

$$\rho_b \qquad\qquad\qquad K_d$$

$$R = 1 + \cfrac{\overbrace{\dfrac{\text{Mass soil}}{\text{Vol aquifer}}} * \overbrace{\dfrac{\text{Mass sorbed compound}}{\text{Mass soil}} \Big/ \dfrac{\text{Mass dissolved compound}}{\text{Vol water}}}}{\underbrace{\text{Vol water} / \text{Vol aquifer}}_{\theta}}$$

$$R = 1 + \frac{\text{Mass sorbed compound}}{\text{Mass dissolved compound}}$$

$$R = \frac{\text{Mass dissolved compound}}{\text{Mass dissolved compound}} + \frac{\text{Mass sorbed compound}}{\text{Mass dissolved compound}}$$

$$R = \frac{\text{Mass total compound (as compound is either dissolved or sorbed)}}{\text{Mass dissolved compound}}$$

This is why the retardation factor has the effect that it does on transport. Consider a contaminant plume consisting of a compound with a retardation factor of 2 in groundwater that is moving with a pore velocity of 1 m/min. Since sorption is assumed to be an infinitely fast process (equilibrium sorption), at any time, half of the total compound mass will be dissolved (and the other half will be sorbed). The dissolved mass is traveling at the pore velocity, and the sorbed mass is not moving. Thus, on average, the plume is moving at the pore velocity divided by the retardation factor, which is what we found, mathematically in Section 2.3.2.1. And of course, for a nonsorbing compound, $R = 1$, meaning all the mass is dissolved, and the compound moves at the pore velocity.

3.6.2 Rate-Limited Sorption

Using our virtual column, let us now examine how sorption kinetics affects concentration distributions in space and time. We will begin by looking at how the magnitude of the first-order desorption rate constant (α) qualitatively and quantitatively affects the distributions.

3.6.2.1 First-Order Kinetics

Figure 3.10 depicts breakthrough curves for our virtual column experiment for three different values of the first-order desorption rate constant. As an aside,

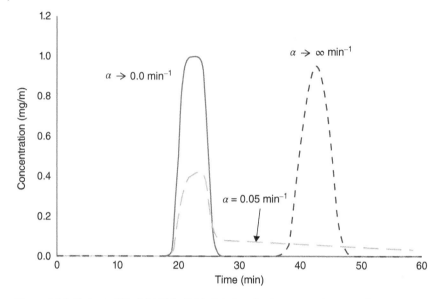

Figure 3.10 Output of Model 106 for Table 3.1 virtual column experiment to demonstrate the effect of sorption kinetics on concentration breakthrough curves at the column outlet. $v = 0.1$ m/min, $\lambda = 0$ min^{-1}, $D = 0.0001$ m^2/min, $k_d = 0.167$ m/kg ($R = 2$).

note that although we are explicitly changing the desorption rate constant, we are also simultaneously changing the sorption rate constant, since the sorption rate constant (k_f) is equal to the desorption rate constant (α) multiplied by the sorption distribution coefficient (k_d) (see Equation (2.7)), and we are maintaining k_d constant.

From Figure 3.10, we see that when $\alpha \to 0$ min^{-1}, the dissolved chemical arrives at the column outlet at the same time as it would if there were no sorption at all. This, of course, makes sense, since a rate constant of zero indicates that the sorption/desorption process is infinitely slow (i.e., it never happens). Compare the $\alpha \to 0$ min^{-1} breakthrough curve in Figure 3.10 (which was simulated using a dispersion coefficient, D, of 0.0001 m^2/min) with the $D = 0.0001$ m^2/min curve in Figure 3.4 (which was simulated assuming no sorption) and we will see the two curves are identical.

Also in Figure 3.10, we see that for instantaneous sorption/desorption ($\alpha \to \infty$ min^{-1}), the breakthrough curve is identical to that for a sorbing chemical, where sorption is modeled as an equilibrium process (compare the $\alpha \to \infty$ min^{-1} breakthrough curve in Figure 3.10 with the $k_d = 0.167$ (mg/kg)/(mg/m) curve in Figure 3.8). Obviously, this makes sense, since if sorption is instantaneous, the process is fast in comparison with other processes affecting transport, so the equilibrium assumption is valid.

Finally, let us look at a moderate sorption/desorption rate, somewhere between infinitely fast and infinitely slow. We choose a value of $\alpha = 0.05\,\text{min}^{-1}$ (more on why we chose this particular value, shortly). We see that for this "moderate" sorption rate, the breakthrough curve is very different from the other, relatively symmetric, breakthrough curves that we have seen. Looking at Figure 3.10, we see that breakthrough for the $\alpha = 0.05\,\text{min}^{-1}$ curve occurs early; at approximately the same time a nonsorbing chemical would break through. And after the initial breakthrough, there is a long tail. What is occurring is that a fraction of the chemical is not sorbed at all, and this fraction breaks through rapidly. However, some of the chemical, which has slowly sorbed, also slowly desorbs, resulting in the long tail. It may not be apparent, as the tail has been cut off, but the area under all three breakthrough curves in Figure 3.10 is the same, as must be the case, since the mass of chemical entering and leaving the column is the same for all three curves. What is not as obvious is that the center of mass (the first moment) of all three breakthrough curves is the same. Indeed, it seems apparent by looking at the curves that the centers of mass are quite different; 22.5 min for the $\alpha \to 0\,\text{min}^{-1}$ curve compared to 42.5 min for the $\alpha \to \infty\,\text{min}^{-1}$ curve. In Chapter 5, where we discuss moments, we will find that in fact the first moments of the three curves are identical, due to a long, albeit too small to distinguish, tail of the $\alpha \to 0\,\text{min}^{-1}$ curve.

Figure 3.11 depicts the spatial distribution of chemical for the three first-order desorption rate constants used in Figure 3.10 ($\alpha \to 0.0$, ∞, and $\alpha = 0.05\,\text{min}^{-1}$). From Figure 3.11, we see that after 15 min the concentration distribution of the chemical that does not sorb ($\alpha \to 0.0\,\text{min}^{-1}$) is symmetric and the front has moved 1.5 m, which makes sense, since the pore velocity, v, is 0.1 m/min. For the chemical that is affected by equilibrium sorption ($\alpha \to \infty\,\text{min}^{-1}$), with a retardation factor, R, of 2, the concentration distribution is also symmetric and the front has moved $\frac{1}{2} * 1.5$ m in 15 min. And for the same reasons as discussed in Figure 3.10, the front of the concentration distribution of chemical that has the moderate desorption rate ($\alpha = 0.05\,\text{min}^{-1}$) has moved 1.5 m, while the distribution is asymmetric with a long tail. Note that the area under the concentration distribution for the instantaneously sorbing chemical ($\alpha \to \infty\,\text{min}^{-1}$) is half the area of the nonsorbing chemical ($\alpha \to 0.0\,\text{min}^{-1}$), since the area represents mass in solution, and for a chemical with $R = 2$, half of the mass is dissolved and half is sorbed. The area under the concentration distribution (and therefore the mass of dissolved chemical) for the slowly sorbing/desorbing chemical is less than the area (and mass) of the nonsorbing chemical distribution but greater than the area (and mass) of the concentration distribution for the instantaneously sorbing chemical.

We can learn something by considering how the concentration distribution of a slowly sorbing/desorbing chemical changes over time. Let us conduct our virtual column experiment for a slowly sorbing/desorbing chemical ($k_d = 0.167\,(\text{mg/kg})/(\text{mg/m})$ so $R = 2$, and $\alpha = 0.05\,\text{min}^{-1}$) and look at the

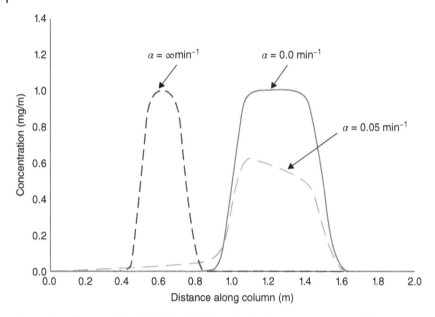

Figure 3.11 Output of Model 106 for Table 3.1 virtual column experiment to demonstrate the effect of sorption kinetics on concentration profiles in space. $t = 15$ min, $v = 0.1$ m/min, $\lambda = 0$ min^{-1}, $D = 0.0001$ m^2/min, $k_d = 0.167$ m/kg ($R = 2$).

concentration distribution of the chemical after 0.5 min (Figure 3.12a). From the figure, we see that the concentration distribution of the slowly sorbing/desorbing chemical is almost identical to the concentration distribution of a nonsorbing chemical. This can be explained by considering the Damköhler number (Da_I), which was discussed in Chapter 2. Recall the Damköhler number is the ratio of an advection timescale to a first-order reaction timescale. In the case of Figure 3.12a, the advection timescale is 0.5 min, and the sorption/desorption first-order reaction timescale is $1/\alpha$ or 20 min. Thus, the Damköhler number is 0.025, indicating the first-order sorption/desorption reaction is 40 times slower than advection, and, therefore, the effect of sorption on the concentration distribution is negligible compared to the effect of advection and can be ignored. Hence, we are able to model the concentration distribution of the slowly sorbing/desorbing chemical after 0.5 min as if it were a nonsorbing chemical, with only advection and dispersion affecting the shape of the distribution.

In Figure 3.12b, we plot the concentration distribution of the slowly sorbing/desorbing chemical after 150 min (note in the figure we have lengthened our virtual column so we can see the entire distribution). In the plot, we compare the distribution to the distribution of a chemical that sorbs instantaneously (equilibrium sorption with $R = 2$). From Figure 3.12b, we see by

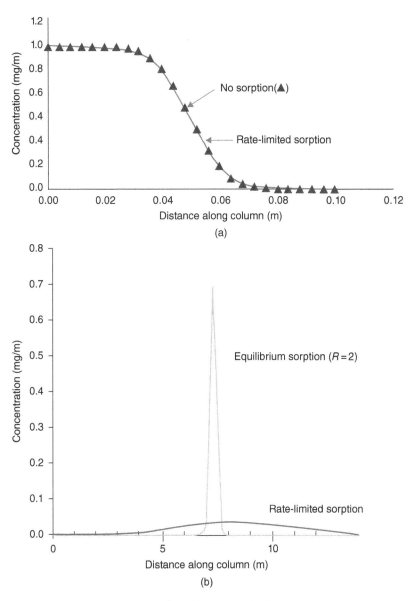

Figure 3.12 Output of Model 106 for Table 3.1 virtual column experiment with $\alpha = 0.05$ min^{-1}, $v = 0.1$ m/min, $\lambda = 0$ min^{-1}, $D = 0.0001$ m^2/min, and $k_d = 0.167$ m/kg ($R = 2$) to demonstrate the effect of sorption kinetics on concentration profiles in space at (a) an early time ($t = 0.5$ min, $Da_l = 0.025$) and (b) a long time ($t = 150$ min, $Da_l = 7.5$).

looking at the areas under the two curves that the dissolved mass of the slowly sorbing chemical appears to be the same as the dissolved mass of the instantaneously sorbing chemical. In fact, using the trapezoidal rule to determine the area under the two curves, we find that both areas represent 0.25 g, which corresponds to half of the mass that was input into our column experiment (1 mg/m * 0.1 m/min * 5 min = 0.5 mg). This is because at equilibrium, with a retardation factor of 2, half of the mass is sorbed and half dissolved. Thus, after 150 min, even our slowly sorbing chemical has reached equilibrium. We can confirm this by calculating the Damköhler number. At 150 min, the Damköhler number is 7.5, indicating advection is 7.5 times slower than the sorption reaction, so that sorption is nearly at equilibrium.

Comparing the distance traveled by the two plumes in Figure 3.12b, we see that the plume with instantaneous sorption has traveled less than the plume of the slowly sorbing/desorbing chemical. Obviously, this is because at early times, the chemical in the slowly sorbing plume was not retarded (per Figure 3.12a), while chemical in the plume where sorption was instantaneous was always retarded. In Chapter 5, when we discuss moment analysis, we quantify the distances traveled (as well as the mass and spread) of plumes that are affected by sorption (both instantaneous and rate-limited).

Finally, we note that the dispersion of the plume with the slowly sorbing chemical is much greater than the spread of the plume where sorption is instantaneous. This is an effect of the rate-limited sorption process, whereby slow sorption/desorption results in enhanced spreading. In Chapter 5, we quantify this effect, and see how an advection/dispersion model, which ignores sorption, can be used to simulate this enhanced spreading at long times (i.e., at large Damköhler numbers). Of course, we have seen that at moderate times, when the Damköhler number is on the order of unity, concentration distributions in space and time are asymmetric with long tails (see the $\alpha = 0.05$ min^{-1} curves in Figures 3.10 and 3.11), and this effect cannot be duplicated by the advection–dispersion model. This is the reason we chose $\alpha = 0.05$ min^{-1} as a convenient desorption rate constant for our column experiment. Since our column length is 2 m and the pore velocity is 0.1 m/min, $Da_I = \frac{L/v}{1/\alpha} = 1.0$, indicating that the advection and sorption reaction timescales are the same, and both processes are important when $\alpha = 0.05$ min^{-1}.

Figure 3.13 demonstrates an interesting phenomenon. For certain combinations of parameter values, rate-limited sorption/desorption can result in dual-peaked concentration profiles in space and time. These dual-peaked profiles occur when desorption is rapid enough to create a second concentration peak that follows the initial peak (which is due to advective transport). This phenomenon only occurs when the desorption rate is "just right." If desorption is slower, the profile is the "typical" long tail profile in time and space as illustrated in Figures 3.10 and 3.11, respectively. If desorption is faster,

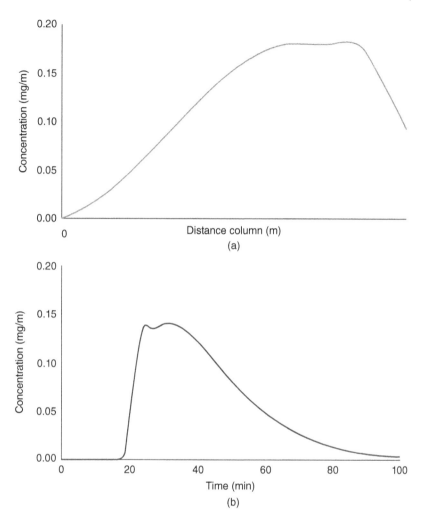

Figure 3.13 Output of Model 106 for Table 3.1 virtual column experiment with $\alpha = 0.15$ min^{-1}, $v = 0.1$ m/min, $\lambda = 0$ min^{-1}, $D = 0.0001$ m^2/min, and $k_d = 0.167$ m/kg ($R = 2$) to demonstrate dual-peaked concentration profiles (a) in space (at time = 21.7 min, $Da_l = 3.3$) and (b) in time ($Da_l = 3$), which result from sorption kinetics.

the profile is the more symmetric profile that is characteristic of equilibrium transport.

3.6.2.2 Diffusion-Limited

As has been demonstrated in a number of papers (Goltz and Roberts, 1988; Parker and Valocchi, 1986; van Genuchten, 1985), the effect of diffusion-limited sorption on transport can be modeled using the same first-order model that was applied in the previous section. To apply these first-order models to simulate diffusive transport and sorption in immobile regions, it is only necessary to convert the diffusion rate constant into an equivalent first-order rate constant.

A diffusion rate constant can be defined as the rate at which compound diffuses through the immobile regions described in Section 2.3.2.2.2. Noting that the diffusion coefficient has units of $[L^2\text{-}T^{-1}]$, we see that a rate constant can be defined by dividing the diffusion coefficient by the square of length scale that is characteristic of the immobile region. Recall in Section 2.3.2.2.2, we defined the parameter "b" as the radius of a spherical or cylindrical immobile region or the half-width of a layered immobile region. Hence, D/b^2 is a rate constant. However, it is necessary to modify this rate constant because sorption **within** the immobile region is assumed to be instantaneous. That is, as the compound diffuses within the immobile region, it is assumed to instantaneously sorb, so the rate of diffusion is retarded by an immobile region retardation factor, R_{im}. As we defined the retardation factor before, $R = 1 + \frac{\rho_b k_d}{\theta}$, where all the parameters (ρ_b, k_d, θ) describe conditions within the immobile region. A number of models have been presented in the literature that allow for different retardation factors in the mobile and immobile regions (e.g., Cameron and Klute, 1977; Rao et al., 1979); for instance, if the immobile region porosity or sorption distribution coefficient differ from those parameters in the mobile region. While these models are more realistic (e.g., to model a sandy mobile region in contact with a clayey immobile region) for the purposes of this text, we will stick with a simplified model that assumes all sorption occurs in the immobile region, and the mobile and immobile region porosity and bulk densities are the same. The interested reader may want to develop equivalent first-order models by using Models 108 and 109, which allow different parameters to be input for immobile and mobile regions.

A diffusion rate constant, modified to account for sorption, is $D/b^2 R$. Using slightly different approaches, Goltz (1986) and van Genuchten (1985) defined first-order rate constants that were equivalent to diffusion rate constants for different immobile region geometries (Table 3.2).

Figure 3.14 demonstrates this equivalence for the Table 3.1 virtual column experiment and assuming diffusion ($D = 6 \times 10^{-8}$ m^2/min) into immobile

Table 3.2 Equivalent first-order rate constants for diffusion into immobile regions of various geometries.

Immobile region geometry	Diffusion rate constant	Equivalent first-order rate constant (Goltz, 1986)	Equivalent first-order rate constant (van Genuchten, 1985)
Spherical (radius $= b$)	$D/b^2 R$	$15 * D/b^2 R$	$22.7 * D/b^2 R$
Cylindrical (radius $= b$)	$D/b^2 R$	$8 * D/b^2 R$	$11.0 * D/b^2 R$
Layered (half-width $= b$)	$D/b^2 R$	$3 * D/b^2 R$	$3.52 * D/b^2 R$

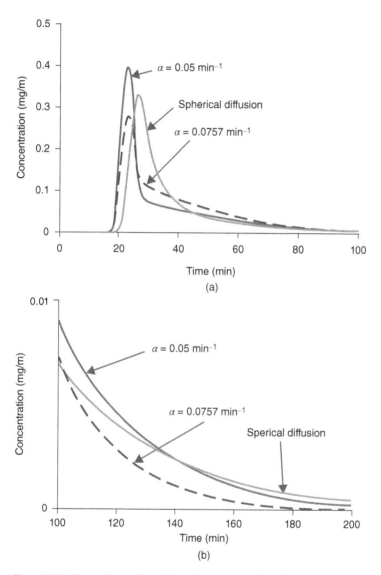

Figure 3.14 Comparison of breakthrough curves simulated assuming sorption may be described by first-order kinetics ($\alpha = 0.05$ and 0.0757 min^{-1}) and by diffusion into spherical immobile regions ($D/(b^2R) = 0.0033$ min^{-1}). Table 3.1 virtual column parameter values with $v = 0.1$ m/min ($q = 0.025$ m/min), $\lambda = 0$ min^{-1}, $D = 0.0001$ m^2/min, and $k_d = 0.167$ m/kg ($R = 2$). First-order sorption kinetics simulated by Model 106. Diffusion into spherical immobile regions simulated using a numerical code to represent Equations (2.23 a), (2.23c), (2.23d), (2.23e), with $v = 3$. Breakthrough curves for (a) $t = 0$–100 min, and (b) $t = 100$–200 min.

regions made up of spheres with radii of 0.003 m. The equivalent first-order rate model simulations come from applying Model 106, with $\alpha = 0.05\,\text{min}^{-1}$ and $\alpha = 0.0757\,\text{min}^{-1}$ for the Goltz (1986) and van Genuchten (1985) equivalence relationships, respectively.

As may be seen in Figure 3.14, the first-order models capture the salient characteristics of the breakthrough curve generated by the more complex diffusion model, particularly the sharp initial breakthrough and the significant tailing. Thus, the observations we made in the prior section with regard to the first-order kinetic model also apply to the diffusion-limited model. However, we also clearly see that there are differences between the first-order and diffusion-limited model simulations. The first-order model simulates sharper leading and trailing edges of the breakthrough curve than is simulated by the diffusion-limited model. This is a consequence of the fact that the first-order model assumes a single concentration for the sorbed (stagnant) chemical. In fact, another way of conceptualizing the first-order model, which is found frequently in the literature, is to replace Equation (2.22c), which describes first-order rate-limited sorption/desorption, with an equation that describes first-order rate-limited transfer of chemical between zones of mobile and immobile water. These two models are mathematically identical and differ only in their conceptual basis. The so-called mobile–immobile model assumes that there is a perfectly mixed immobile region into which chemical is transferred, by a first-order rate process, to and from the mobile region. The assumption that the immobile zone is perfectly mixed means that the concentration gradient between the two zones is maximized, so mass transfer of chemical between the two zones is also maximized. This results in the breakthrough curves observed in Figure 3.14a for the first-order models being sharper than the breakthrough curve simulated by the diffusion-limited model, where concentrations in the immobile region are a function of position within the immobile region. In the diffusion-limited model, concentration gradients are less and mass transfer into and out of the immobile region is slower than simulated by the mobile–immobile model. Ultimately, at long times (Figure 3.14b), we note that the breakthrough curve tailing simulated by the diffusion-limited model is more extensive than tailing simulated by the mobile–immobile model, due to the slower transfer of chemical out of the immobile region in the diffusion-limited model simulation.

3.7 Effect of First-Order Degradation

Figures 3.15 and 3.16 show the effect of first-order degradation on concentration versus time breakthrough curves and concentration versus distance profiles, respectively. We see that at high values of the Damköhler number, when the advection timescale (L/v or t) is considerably longer than the first-order

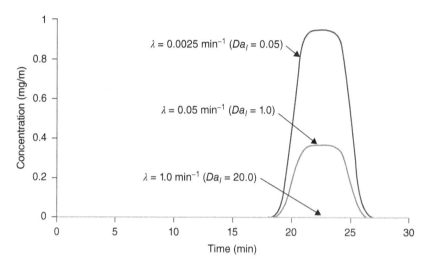

Figure 3.15 Output of Model 106 for Table 3.1 virtual column experiment to demonstrate the effect of first-order degradation on concentration breakthrough curves at the column outlet. $v = 0.1$ m/min, $D = 0.0001$ m²/min.

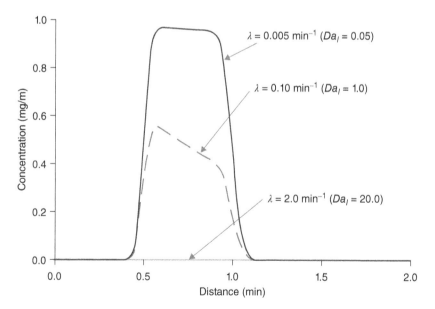

Figure 3.16 Output of Model 106 for Table 3.1 virtual column experiment to demonstrate the effect of first-order degradation on concentration profiles in space. $t = 10$ min, $v = 0.1$ m/min, $D = 0.0001$ m²/min.

degradation reaction timescale $(1/\lambda)$, degradation dominates, and concentrations are near zero. Conversely, at small Damköhler numbers, when the advection timescale is considerably shorter than the reaction timescale, advection dominates and the concentration distributions in time and space are similar to what they would be if no degradation were occurring. It is only when the advection and degradation reaction timescales are similar $(Da_I \sim 1)$ that the effect of both advection and degradation on the concentration distributions in space and time may be clearly observed. In Figure 3.15, we see that the breakthrough curve for $Da_I = 1.0$ is relatively symmetric, while from Figure 3.16 we note the $Da_I = 1.0$ concentration distribution is quite asymmetric. This is explained by the fact that for the breakthrough curve, on average compound molecules in both the leading and lagging portions of the curve have been in the column (and therefore, subject to degradation) for the same amount of time, whereas for the concentration profile in space, compound molecules represented in the leading portion of the profile have been in the column for time t, while molecules in the lagging portion of the profile have been in the column for time $t - t_s$. Hence, concentrations in the leading portion of the profile are less than the concentrations in the lagging portion of the profile, since the compound in the leading part of the profile has had more time to degrade.

Figures 3.17 and 3.18 show the interaction between sorption and first-order degradation on concentration versus time breakthrough curves and

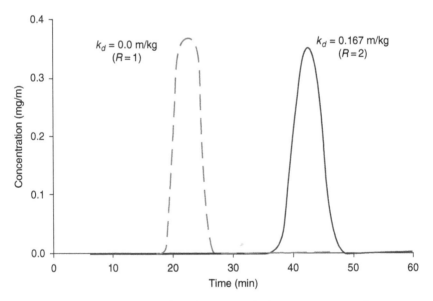

Figure 3.17 Output of Model 106 for Table 3.1 virtual column experiment to demonstrate the effect of sorption on the concentration breakthrough curves at the column outlet of a degrading compound. $v = 0.1$ m/min, $D = 0.0001$ m²/min, $\lambda = 0.05$ min⁻¹ $(Da_I = 1)$, $\alpha = $ INF.

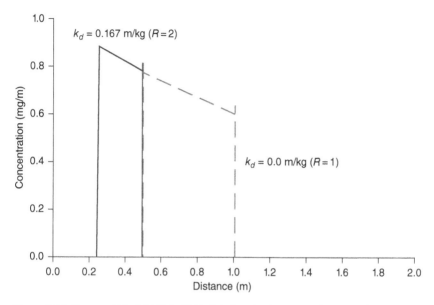

Figure 3.18 Output of Model 106 for Table 3.1 virtual column experiment to demonstrate the effect of sorption on concentration profiles in space of a degrading compound. $t = 10$ min, $v = 0.1$ m/min, $D = 0.0001$ m^2/min, $\lambda = 0.05$ min^{-1} $(Da_I = 0.5)$, $\alpha = $ INF.

concentration versus distance profiles, respectively. From Figure 3.17, we see that the areas under both retarded and nonretarded breakthrough curves are the same, meaning that the mass of compound exiting the column is the same in both cases. We may wonder why this is so, since obviously the retarded compound is in the column longer, and therefore subject to degradation for a longer period of time. However, recall our assumption that degradation only occurs in the dissolved phase. Thus, for a compound with retardation factor, R, even though the compound is degrading for a period that is R times longer than a nonretarded compound, the mass of compound that is subject to this degradation is R times less. Thus, overall, the mass of compound that leaves the column is the same for both retarded and nonretarded compounds.

From Figure 3.18, we see that the areas under the concentration profiles in space (and therefore the masses) of a retarded and nonretarded compound are obviously different. This difference in mass is partly explained by the explanation given for Figure 3.9. That is, the profiles only show dissolved compound, and since $R = 2$, half of the mass of the retarded compound is sorbed. Therefore, we would expect the area under the retarded compound profile to be half the area of the nonretarded compound's profile, as seen in Figure 3.9. However, we also need to account for degradation. As explained earlier, only the dissolved portion of the retarded compound is subject to degradation, while the entire nonretarded compound is degrading (since the nonretarded compound is completely in the dissolved phase). Thus, the area under the retarded compound's profile is, in fact, greater than half of the area of the nonretarded compound's profile. We examine this effect quantitatively in Chapter 5.

3.8 Effect of Boundary Conditions

3.8.1 Effect of Boundary Conditions on Breakthrough Curves

We now look at the effect on concentration profiles of using either a first-type or third-type boundary condition at $x = 0$. Recall a first-type BC was (Equation (2.29)) defined in Chapter 2 as

$$C(x = 0, t) = C_0$$

and a third-type BC (Equation (2.31a)) was defined as

$$vC(x = 0, t) - D_x \frac{dC(x = 0, t)}{dx} = vC_0$$

We can see the effect of BC on the breakthrough concentration curve by comparing the output of Model 102 (first-type BC) with Model 106 (third-type BC) for the virtual experimental column described in Table 3.1.We see that for the conditions simulated in Figure 3.19, the BC at $x = 0$ has no effect on the breakthrough curve at the column outlet. This is due to the Peclet number used for the simulation. Note that Figure 3.19 was constructed for a Peclet number (*Pe*) of 2000, meaning that the effect of dispersion is very small. To see how dispersion and the BC are related, multiply both sides of Equation (2.29) by the pore velocity, v, to obtain Equation (3.12):

$$vC(x = 0, t) = vC_0 \tag{3.12}$$

Comparing Equation (3.12) for a first-type BC and Equation (2.31a) [that follows] for a third-type BC,

$$vC(x = 0, t) - D_x \frac{dC(x = 0, t)}{dx} = vC_0$$

we see that the only difference between the two BCs is due to dispersion. That is, a first-type BC, expressed as Equation (3.12), assumes flux due to advection is constant, while a third-type BC (Equation (2.31a)) assumes total flux (flux due to advection plus flux due to dispersion) is constant. Thus, if dispersive flux is negligible when compared to advective flux, as is the case when the Peclet number is large, there is no difference in the concentration profiles simulated assuming first-type or third-type BCs. On the other hand, Figure 3.20 shows that when dispersion is significant when compared to advection ($Pe = 4$), there is a significant difference between the breakthrough concentrations simulated using a first- and third-type BC.

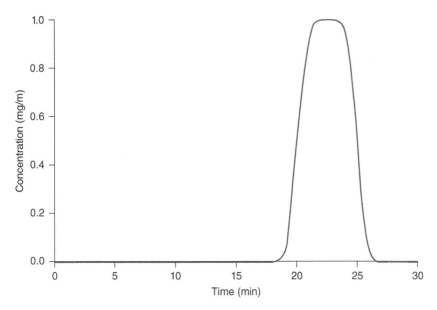

Figure 3.19 Output of Models 102 (first-type BC at $x = 0$) and 106 (third-type BC at $x = 0$) for Table 3.1 virtual column experiment to demonstrate the effect of BC on the concentration breakthrough curves at the column outlet at high *Pe*. $v = 0.1$ m/min, $\lambda = 0$ min^{-1}, $k_d = 0$ m/kg, $D = 0.0001$ m^2/min ($Pe = 2000$).

The relative behavior of the two breakthrough curves shown in Figure 3.20 can be understood by comparing Equations (2.31a) and (3.12). Equation (2.31a) indicates that total flux is constant, while Equation (3.12) indicates that advective flux is equal to the same constant. Since total flux is the sum of advective and dispersive fluxes, the simulated concentration, C, at $x = 0$, for a third-type BC (let us designate this as C_{III}) will be less than the simulated concentration at $x = 0$ for a first-type BC (C_I). Mathematically, we can see this by subtracting Equation (3.12) from Equation (2.31a) and rearranging terms to obtain

$$v[C_I(x = 0, t) - C_{III}(x = 0, t)] = -D_x \frac{dC_{III}(x = 0, t)}{dx} \tag{3.13}$$

Since the term on the right-hand side of Equation (3.13) representing dispersive flux is positive (concentration decreases in the positive x-direction, since the maximum concentration is at $x = 0$), C_I must be greater than C_{III}. Hence, the breakthrough curve simulated assuming a first-type BC at $x = 0$ has a higher concentration peak than the breakthrough curve that was simulated assuming a third-type BC at $x = 0$. Also note that when the chemical input at $x = 0$ is "turned off" (i.e., when time, t, is greater than the pulse duration, t_s, C $(x = 0, t) = 0$) and the concentration gradient at $x = 0$ becomes positive in the x-direction (concentrations increase with x, as the concentration at $x = 0^+$ is greater than

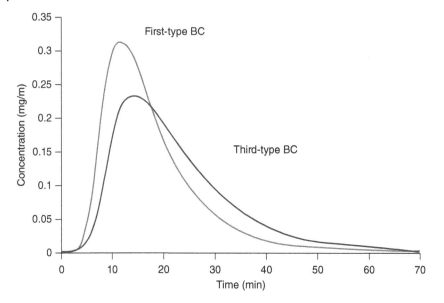

Figure 3.20 Output of Models 102 (first-type BC at $x = 0$) and 106 (third-type BC at $x = 0$) for Table 3.1 virtual column experiment to demonstrate the effect of BC on the concentration breakthrough curves at the column outlet at low *Pe*. $v = 0.1$ m/min, $\lambda = 0$ min^{-1}, $k_d = 0$ m/kg, $D = 0.05$ m^2/min ($Pe = 4$).

the concentration at $x = 0$), so the dispersive flux on the right-hand side of Equation (3.13) becomes negative. Under these conditions, C_{III} is greater than C_I. Hence, after the breakthrough curves peak, we begin to see that a third-type BC simulates higher concentrations than are simulated with a first-type BC.

As would be expected, the areas under the two breakthrough curves are the same, since the total flux of chemical entering the column is the same for both BCs (vC_0), and the chemical pulse duration (t_s) is the same as well.

3.8.2 Volume-Averaged Resident Concentration Versus Flux-Averaged Concentration

Though perhaps not immediately obvious, the above discussion on BCs leads into a discussion regarding different ways that concentration might be measured. Thus far in this text, we have implicitly assumed that concentration is measured by instantaneously taking a volume of stagnant water and measuring the mass of compound present in that volume. This "volume-averaged" concentration has also been called "resident" concentration. However, another way of measuring concentration might be, for example, to accumulate the water flowing out of an experimental column over a short time interval and then measure the mass of chemical in that volume of water. This measurement gives us a flux-averaged concentration. The flux-averaged concentration is the mass flux of chemical (J, defined in Equation (2.2)) divided by the specific discharge or flux of water ($q = \theta v$). Since J is the sum of advective and dispersive fluxes, we may write

$$J = \theta v C_r - \theta D \frac{\partial C_r}{\partial x} \qquad (3.14)$$

where C_r is the resident concentration. Since the flux-averaged concentration, C_f, is defined as mass flux divided by specific discharge:

$$C_f = \frac{J}{q} \qquad (3.15)$$

We can substitute Equation (3.14) into Equation (3.15) to obtain the relationship between flux-averaged and resident concentration:

$$C_f = C_r - \frac{D}{v} \frac{\partial C_r}{\partial x} \qquad (3.16)$$

Using this relationship, it is easy to show (e.g., Kreft and Zuber, 1978; van Genuchten and Parker, 1984; Schwartz et al., 1999) that the ADR is the same whether the dependent variable is flux-averaged or volume-averaged concentration. The only difference is that if a third-type BC is used for the ADR written in terms of a volume-averaged concentration, a first-type BC should be used for the ADR written in terms of a flux-averaged concentration. This is readily seen by writing the third-type BC equation (Equation (2.31a)) in terms of resident concentration:

$$C_r(x = 0, t) - \frac{D}{v} \frac{\partial C_r(x = 0, t)}{\partial x} = C_0 \qquad (3.17)$$

and then using Equation (3.16) to substitute C_f for the left-hand side, to obtain

$$C_f(x = 0, t) = C_0 \qquad (3.18)$$

Thus, looking at Figure 3.20, the third-type BC breakthrough curve can be thought of as the breakthrough curve that would be observed if we measured resident concentration at the column outlet, while the first-type BC breakthrough curve can be thought of as the curve that would be observed if we measured flux-averaged concentrations at the outlet. van Genuchten and Parker (1984) recommend that for a column experiment, where the breakthrough curve is measured at the outlet of the column, it is most appropriate to assume flux-averaged concentrations are being measured, so that a model with a first-type BC at the column inlet should be used. Of course, as we see when comparing Figures 3.19 and 3.20, there is only a significant difference between the curves simulated assuming a first-type and third-type BC when the Peclet number is low (e.g., Schwartz et al. (1999) state that the difference is significant when $Pe < 10$).

Problems

3.1 Use AnaModelTool Model 106 to reproduce Figures 3.2 and 3.3. Why is it possible to reproduce the two figures, which were simulated using Model 102, by running Model 106, which has different boundary conditions?

3.2 Compare concentration versus time simulations at $x = 2$ m for Models 106 (finite column length) and 107 (semi-infinite column). Use the parameter values in Table 3.1 with pore velocity $(v) = 0.1$ m/min, longitudinal dispersion $(D_x) = 0.01$ m^2/min, sorption distribution coefficient $(k_d) = 0.1667$ m/kg, equilibrium sorption and no degradation. For Model 106, what value of the column length (L) gives results that are equal to the semi-infinite model? What criteria did you use to decide that the breakthrough curves simulated by the two models were equal?

3.3 Compare concentration versus time simulations at $x = 2$ m for Models 103 (first-type boundary condition at $x = 0$) and 107 (third-type boundary condition at $x = 0$) for different values of longitudinal dispersion (D_x). Use the parameter values in Table 3.1 with pore velocity $(v) = 0.1$ m/min, no sorption and no degradation. What is the maximum value of the longitudinal dispersion coefficient at which the results of the two models equal? What is the Peclet number at this value of longitudinal dispersion? What criteria did you use to decide that the breakthrough curves simulated by the two models were equal?

References

Cameron, D.R. and A. Klute, Convective–dispersive solute transport with a combined equilibrium and kinetic adsorption model, *Water Resources Research*, 13(1): 183–188, 1977.

de Hoog, F.R., Knight, J.H., and Stokes, A.N., An improved method for numerical inversion of Laplace transforms. *SIAM Journal on Scientific and Statistical Computing*, 3, 357–366, 1982.

Goltz, M.N., Three-dimensional analytical modeling of diffusion-limited solute transport, Ph.D. Thesis, Stanford University, Stanford, CA, 172 pp., 1986.

Goltz, M.N. and P.V. Roberts, Simulations of physical nonequilibrium solute transport models: application to a large-scale field experiment, *Journal of Contaminant Hydrology*, 3, 37–63, 1988.

Hollenbeck, K.J., INVLAP.M: A MATLAB® function for numerical inversion of Laplace transforms by the de Hoog algorithm, https://www.mathworks.com/matlabcentral/answers/uploaded_files/1034/invlap.m, 1998.

Kreft, A. and A. Zuber, On the physical meaning of the dispersion equation and its solutions for different initial and boundary conditions, *Chemical Engineering Science, 33*(11): 1471–1480, 1978.

Parker, J.C. and A.J. Valocchi, Constraints on the validity of equilibrium and first-order kinetic transport models in structured soils, *Water Resources Research, 22*(3): 399–407, 1986.

Rao, P.S.C., J.M. Davidson, R.E. Jessup, and H.M. Selim, Evaluation of conceptual models for describing nonequilibrium adsorption-desorption of pesticides during steady-flow in soils, *Soil Science Society of American Journal, 43*, 22–28, 1979.

Schwartz, R.C., K.J. McInnes, A.S.R. Juo, L.P. Wilding, and D.L. Reddell, Boundary effects on solute transport in finite soil columns, *Water Resources Research, 35*(3): 671–681, 1999.

van Genuchten, M.Th., A general approach for modeling solute transport in structured soils, *International Association of Hydrogeologists: Memoires, 17*, 513–526, 1985.

van Genuchten, M.Th., and J.C. Parker, Boundary conditions for displacement experiments through short laboratory soil columns, *Soil Science Society of American Journal, 48*, 703–708, 1984.

4

Analytical Solutions to 3-D Equations

4.1 Solving the ADR Equation with Initial/Boundary Conditions

We can solve the three-dimensional advective–dispersive–reactive (ADR) equation (Equation (2.21b)) following the method shown in Appendices A and B for the one-dimensional ADR. The key difference is that now we must also account for the extra spatial dimensions. We do this by using Fourier transforms in space, in addition to taking the Laplace transform in time. This approach is applied in Appendices F and G for an instantaneous Dirac point source at $x = y = z = 0$. The Appendix F solution is for Equation (2.21b) with first-order degradation kinetics, while the Appendix G solution is for Equation (2.21b) with zeroth-order degradation kinetics. The initial condition for an instantaneous point source of mass M with sorption described by an equilibrium, linear isotherm (retardation factor R) is $C(x, y, z, t = 0) = \frac{M}{R}\delta(x)\delta(y)\delta(z)$. Boundary conditions are set at $C(x, y, z, t) = 0$ at $x = y = z = \infty$. As shown in Appendix F, the solution of the PDE with its IC/BCs for first-order degradation kinetics is

$$C(x, y, z, t) = \frac{M\sqrt{R}\exp\left[-\dfrac{R(x - v/_R t)^2}{4D_x t} - \dfrac{Ry^2}{4D_y t} - \dfrac{Rz^2}{4D_z t} - \dfrac{\lambda}{R}t\right]}{(4\pi t)^{3/2}\sqrt{D_x D_y D_z}} \tag{4.1}$$

while Appendix B has the solution for zeroth-order degradation kinetics:

$$C(x, y, z, t) = \frac{M\sqrt{R}\exp\left[-\dfrac{R(x - v/_R t)^2}{4D_x t} - \dfrac{Ry^2}{4D_y t} - \dfrac{Rz^2}{4D_z t}\right]}{(4\pi t)^{3/2}\sqrt{D_x D_y D_z}} - \frac{k_0}{R}t \tag{4.2}$$

Note that due to the second term on the right-hand side of Equation (4.2), the solution for the zeroth-order kinetic model can potentially result in negative

Analytical Modeling of Solute Transport in Groundwater: Using Models to Understand the Effect of Natural Processes on Contaminant Fate and Transport, First Edition. Mark Goltz and Junqi Huang.
© 2017 John Wiley & Sons, Inc. Published 2017 by John Wiley & Sons, Inc.
Companion Website: www.wiley.com/go/Goltz/solute_transport_in_groundwater

values of concentration – clearly, a nonsensical result that indicates a limitation of the zeroth-order model.

Appendix H provides citations for where additional analytical solutions may be found in the literature for various initial and boundary conditions.

4.2 Using Superposition to Derive Additional Solutions

As we saw in Chapter 3, since the ADR equation is a linear PDE (so long as the expressions for sorption and reaction are also linear), solutions can be super-posed (also see Section 1.3.5). The AnaModelTool software includes solutions for various IC/BCs (Appendix H). In subsequent sections, we use these solutions to illustrate the effect of processes on three-dimensional ADR transport.

4.3 Virtual Experimental System

Let us "build" a virtual three-dimensional experimental system of length l, width w, and height h, in the x-, y-, and z-directions, respectively (Figure 4.1). The system is initially uncontaminated, with a rectangular source of width y_t-y_b and height z_t-z_b at a concentration C_0 at the $x = 0$ boundary. Groundwater flow is in the positive x-direction. The source is "active" for a time period t_s, after which the source concentration goes to zero. Let us assume the system is filled with a porous material that has porosity, θ, and bulk density, ρ_b. Parameter values for our virtual system are listed in Table 4.1.

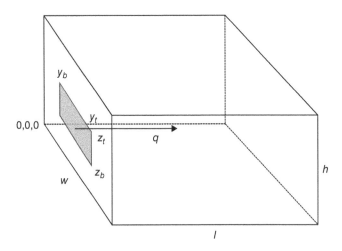

Figure 4.1 Three-dimensional virtual aquifer system.

Table 4.1 Three-dimensional virtual system characteristics.

Parameter	Value
Length (l), width (w), and height (h)	$l = 20$ m $w = 10$ m $h = 6$ m
Source width ($y_t - y_b$) and height ($z_t - z_b$)	$y_t - y_b = 4$ m ($y_t = 7$ m, $y_b = 3$ m) $z_t - z_b = 2$ m ($z_t = 4$ m, $z_b = 2$ m)
Source concentration (C_0)	1 mg/L
Source duration (t_s)	5 min
Porous media porosity (θ)	0.25
Porous media bulk density (ρ_b)	1.5 kg/L
Groundwater pore velocity (v)	0.1 m/min

In the following sections, we use AnaModelTool to perform "experiments" in our virtual system. We vary the individual parameter values in our model equations to observe the effect of different processes on the simulated chemical concentrations as a function of space and time.

4.4 Effect of Dispersion

Let us use Model 301 in AnaModelTool to investigate the effect of dispersion in a three-dimensional system. To begin, we will minimize dispersion in the y- and z-directions by setting D_y and D_z to low values (0.0002 m²/min). Degradation and sorption are also turned off ($\lambda = k_d = 0$). We now run simulations where the longitudinal dispersion coefficient, D_x, is varied to investigate the maximum concentration encountered at a sampling well located 1 m downgradient of the center of the rectangular source described in Table 4.1 (sampling well at $x = 1$ m, $y = 5$ m, and $z = 3$ m). Figure 4.2 is a plot of maximum concentration at the sampling well as a function of the Peclet number (Pe), where the Peclet number is defined, as in Chapters 2 and 3, as the dimensionless ratio of a dispersion timescale (L^2/D_x) and an advection timescale (L/v), where $L = 1$ m, the distance from the source to the sampling well. Thus, $Pe = \frac{Lv}{D_x} = \frac{0.1\,\mathrm{m^2/min}}{D_x}$ for the parameters in this problem.

Figure 4.2 exhibits some interesting behavior. We see at very high Pe number ($Pe > 1000$), where longitudinal dispersion is minimal, the maximum concentration observed at the monitoring well is equal to the input concentration

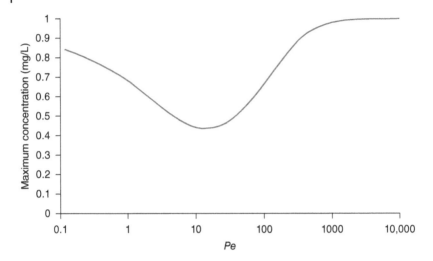

Figure 4.2 Output of Model 301 for the virtual three-dimensional system (Table 4.1) to demonstrate the effect of longitudinal dispersion on the maximum concentration observed at a monitoring well 1 m downgradient of the rectangular source ($x = 1$ m, $y = 5$ m, $z = 3$ m). $D_y = D_z = 0.0002$ m^2/min, $\lambda = 0$ min^{-1}, $k_d = 0$ L/kg, $\alpha = $ INF.

(since dispersion in the y- and z-directions is also negligible and there is no degradation). As the Pe number decreases, the impact of longitudinal dispersion becomes more important and the maximum concentration observed at the monitoring well decreases due to longitudinal spreading. Interestingly, though, we see from Figure 4.2 that at about $Pe = 10$, the observed maximum concentration starts to increase as the Pe number continues to decrease. What is happening at these low values of Pe is that the mass transport of chemical from the source to the monitoring well due to dispersion is significant compared to the mass transport due to advection, and this additional mass reaching the well at short times results in a higher maximum concentration at the well.

Figure 4.3 shows spatial concentration profiles in (a) plan and (b) cross section for "typical" values of the dispersion coefficient. As a rule of thumb, transverse dispersivity is 1/10th the value of longitudinal dispersivity, and vertical dispersivity is quite small. For Figure 4.3, dispersivities of 1.0, 0.1, and 0.002 m were chosen for the longitudinal, transverse, and vertical directions, respectively (resulting in dispersion coefficient values of 0.1, 0.01, and 0.0002 m^2/min, respectively, since the groundwater pore velocity, v, is 0.1 m/min). In plan, the isoconcentration lines that depict the contaminant plume form relatively symmetric ellipses (except for the plume leading edge, which is flattened a bit due to the first-type boundary condition at $x = 20$ m). Due to the relative values of dispersivity in the three directions, the plume is longest in the x-direction, with less spreading in the y-direction and considerably less spreading in the z-direction. One interesting aspect of the simulation

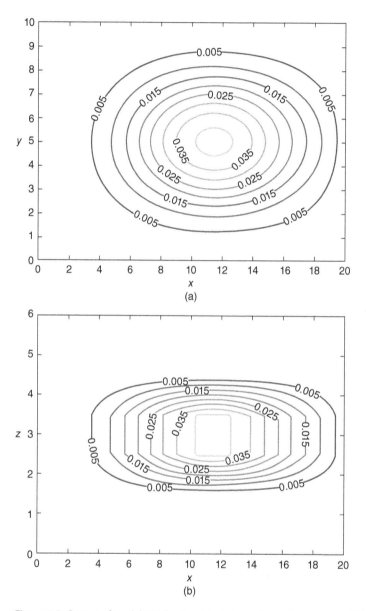

Figure 4.3 Output of Model 301 for the virtual three-dimensional system (Table 4.1) to demonstrate how "typical" ratios of longitudinal to transverse dispersion and longitudinal to vertical dispersion affect the morphology of a contaminant plume. $D_x = 0.1$ m^2/min, $D_y = 0.01$ m^2/min, $D_z = 0.0002$ m^2/min, $\lambda = 0$ min^{-1}, $k_d = 0$ L/kg, $\alpha = $ INF. (a) Plan view (at $z = 3$ m) and (b) vertical cross section (at $y = 5$ m) at $t = 100$ min. All coordinates in meters and note the axes have different scales.

shown in Figure 4.3 is the location of the center of mass. Just eyeballing the plume, it appears that the center of mass is at coordinates ($x = 11.5$ m, $y = 5$ m, $z = 3$ m). The location of the center of mass in the y- and z-directions is no surprise, since the center of the source zone is at those same y- and z-coordinates, and there is no net flux in those directions, due either to advection or dispersion, that would serve to displace the center of mass location. In the x-direction, though, we see the center of mass is at $x = 11.5$ m. Due to advection alone, we would expect the center to be at $x = 10$ m, since the pore velocity is 0.1 m/min and the spatial concentration profile is simulated at 100 min. In space, since dispersion acts in both positive and negative directions, we might expect that there is no net effect of dispersion on the location of the center of mass, so the center should still be at $x = 10$ m after we account for dispersion. Obviously, looking at Figure 4.3, we see that is not the case. So what is going on?

Let us run a few more simulations to try to determine the reason the center of mass in Figure 4.3 moved further in the x-direction than predicted by advection alone. Figure 4.4 depicts simulation results of Model 306, which is for an instantaneous Dirac pulse source at the origin of an infinite system. The dispersion coefficients and pore velocity are the same as used in Figure 4.3, and the spatial concentration profile is again simulated at 100 min. In Figure 4.4, the center of mass is where we expect it to be, at coordinates $x = 10.0$ m, $y = 0$ m, $z = 0$ m. The center of mass was displaced due to advection alone ($v = 0.1$ m/min for 100 min) with dispersion playing no role, since dispersion acted in both the positive and negative x-directions equally.

The difference in the results shown in Figures 4.3 and 4.4 is the boundary condition, specifically, the first-type boundary condition used in Model 301 at the $x = 0$ m boundary. Let us consider what happens right next to the $x = 0$ m boundary as contaminant advects in the positive x-direction. Both advection and dispersion act in the positive x-direction, but due to the boundary condition, dispersion cannot act in the negative x-direction. This is in contrast to the infinite boundary conditions simulated in Figure 4.4, where dispersion acts equally in both positive and negative directions. Due to this asymmetry in how dispersion acts at the boundary, the center of mass moves further in the x-direction than predicted due to advection alone. Figure 4.5 shows that for a first-type boundary condition at $x = 0$ m, for low values of dispersion (corresponding to high values of the Peclet number) the movement of the center of mass is controlled by advection alone, and the center of mass is at 10 m ($v = 0.1$ m/min for 100 min). As dispersion increases, and the Peclet number decreases, the impact of dispersion at the boundary becomes significant and the x-coordinate of the center of mass monotonically increases.

The effect of dispersion in a three-dimensional system on the behavior of concentration versus time breakthrough curves is as would be expected. Figure 4.6 shows simulated concentration versus time data for two sampling wells. One sampling well (breakthrough curve depicted in Figure 4.6a) is

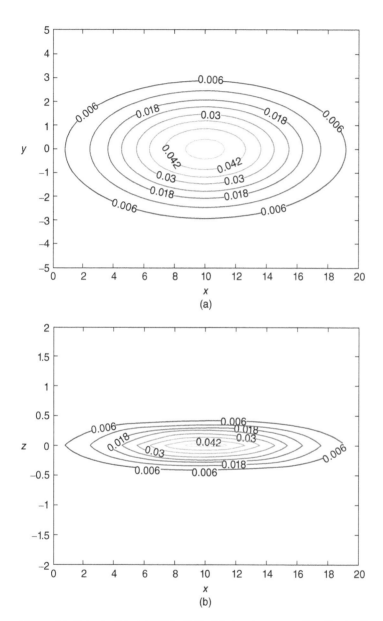

Figure 4.4 Output of Model 306 for Table 4.1 virtual system soil and hydraulic characteristics ($\theta = 0.25$, $\rho_b = 1.5$ kg/L, $v = 0.1$ m/min) and $M = 1$ mg, $D_x = 0.1$ m^2/min, $D_y = 0.01$ m^2/min, $D_z = 0.0002$ m^2/min, $\lambda = 0$ min^{-1}, $k_d = 0$ L/kg, $\alpha =$ INF. (a) Plan view (at $z = 0$ m) and (b) vertical cross section (at $y = 0$ m) at $t = 100$ min. All coordinates in meters and note the axes have different scales.

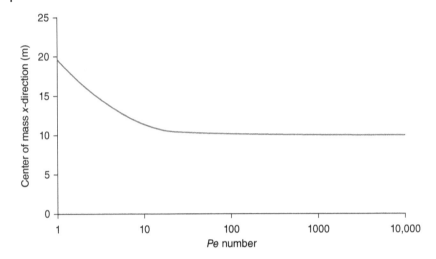

Figure 4.5 Approximate location of the center of mass in the x-direction as a function of dispersion in the x-direction, D_x (log scale), simulated by Model 304 for Table 4.1 virtual system soil and hydraulic characteristics ($\theta = 0.25$, $\rho_b = 1.5$ kg/L, $v = 0.1$ m/min) and $D_y = 0.01$ m²/min, $D_z = 0.0002$ m²/min, $\lambda = 0$ min^{-1}, $k_d = 0$ L/kg, $\alpha = $ INF, $t = 100$ min.

at a location downgradient from, but within, the rectangular source y–z coordinates ($x = 10$ m, $y = 3.1$ m, $z = 3$ m), while the second sampling well (breakthrough curve depicted in Figure 4.6b) is at a location downgradient from, but outside of, the rectangular source y–z coordinates ($x = 10$ m, $y = 2.9$ m, $z = 3$ m). Figure 4.6a shows that for a well downgradient from, but within, the rectangular source y–z coordinates, as the value of the dispersion coefficient in the y-direction increases, the breakthrough concentrations decrease, while the opposite is true for a well downgradient from, but outside of, the rectangular source y–z coordinates. Such a behavior is anticipated, since increased transverse dispersion will serve to more rapidly decrease concentrations being transported from a source (Figure 4.6a), at the same time, wells not directly downgradient from the source, which would otherwise be minimally impacted by the concentration plume, begin to see significant chemical concentrations as transverse dispersion increases (Figure 4.6b).

4.5 Effect of Sorption

4.5.1 Linear, Equilibrium Sorption

As we saw in Chapters 2 and 3, the effect of sorption is to slow the advection and dispersion processes by a factor, R, the retardation factor. Figure 4.7 clearly shows this same effect of sorption in three dimensions. Figures 4.7a and b

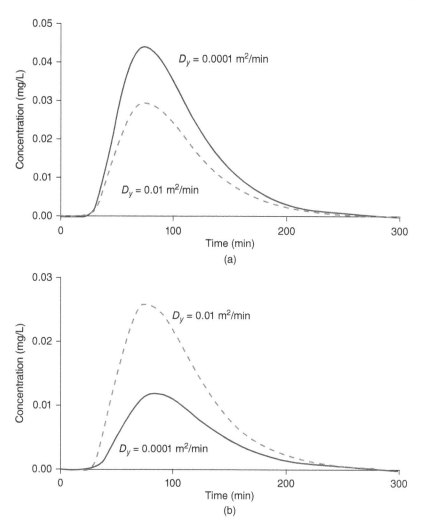

Figure 4.6 Concentration versus time output of Model 304 for Table 4.1 virtual system characteristics to demonstrate effect of transverse dispersion in a three-dimensional system. $D_x = 0.1$ m²/min, $D_y = 0.01$ or 0.0001 m²/min, $D_z = 0.0002$ m²/min, $\lambda = 0$ min⁻¹, $k_d = 0$ L/kg, $\alpha = $ INF. (a) Sampling well at $x = 10$ m, $y = 3.1$ m, $z = 3$ m, and (b) sampling well at $x = 10$ m, $y = 2.9$ m, $z = 3$ m.

are concentration profiles in space for a nonretarded, conservative chemical ($R = 1$) at $t = 100$ min, while Figures 4.7c and d are concentration profiles in space for a retarded chemical ($R = 2$) at $t = 200$ min. We see that the center of mass of the conservative chemical has moved 10 m in 100 min, since the groundwater velocity is 0.1 m/min. The center of mass of the retarded chemical has moved the same distance in twice the time, since advection is retarded by

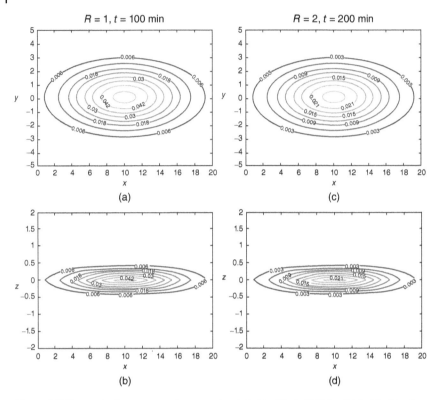

Figure 4.7 Concentration contours in space simulated by Model 306 for Table 4.1 virtual system soil and hydraulic characteristics ($\theta = 0.25$, $\rho_b = 1.5$ kg/L, $v = 0.1$ m/min) and $M = 1$ mg, $D_x = 0.1$ m^2/min, $D_y = 0.01$ m^2/min, $D_z = 0.0002$ m^2/min, $\lambda = 0$ min^{-1}, $\alpha =$ INF. (a) Plan view (at $z = 0$ m) for $k_d = 0.0$ L/kg ($R = 1$) at $t = 100$ min, (b) vertical cross section (at $y = 0$ m) for $k_d = 0.0$ L/kg ($R = 1$) at $t = 100$ min, (c) plan view (at $z = 0$ m) for $k_d = 0.167$ L/kg ($R = 2$) at $t = 200$ min, and (d) vertical cross section (at $y = 0$ m) for $k_d = 0.167$ L/kg ($R = 2$) at $t = 200$ min. All coordinates in meters and note the axes have different scales.

a factor of 2. Similarly, we see that the spread of the retarded chemical after 200 min is identical to the spread of the conservative chemical after 100 min. The only difference between the retarded and conservative chemical contours is the values of the isoconcentration contours. The concentration values for the conservative chemical are twice as high as for the retarded chemical, since half the mass of the retarded chemical is sorbed.

4.5.2 Rate-Limited Sorption

Figures 4.8a and b show plan views of concentration contours in space for a sorbing compound, where sorption is instantaneous and rate limited, respectively. We observe several important differences between the two

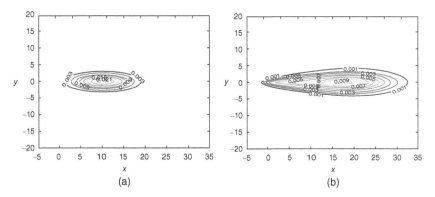

Figure 4.8 Plan view at $z = 0$ m, $t = 200$ min, of concentration contours in space simulated by Model 306 for Table 4.1 virtual system soil and hydraulic characteristics ($\theta = 0.25$, $\rho_b = 1.5$ kg/L, $v = 0.1$ m/min) and $M = 1$ mg, $D_x = 0.1$ m^2/min, $D_y = 0.01$ m^2/min, $D_z = 0.0002$ m^2/min, $\lambda = 0$ min^{-1}, $k_d = 0.167$ L/kg ($R = 2$). (a) $\alpha = $ INF, (b) $\alpha = 0.005$ min^{-1}. All coordinates in meters.

figures. First, we see that the peak of the plume has moved further in the case where sorption is modeled as a rate-limited process. This is expected, since due to the slower sorption process, a fraction of dissolved chemical is less affected by sorption and therefore is able to move through the system faster (i.e., retarded less) than when sorption is modeled as an instantaneous process.

Second, we see that where sorption is modeled as instantaneous, the plume is symmetric, whereas for rate-limited sorption, the plume is asymmetric, with a sharp front and long tail. Finally, we note that for the rate-limited sorption simulation, upgradient (at $x < 0$) and cross-gradient (y-values at $x = 0$ m) concentrations are significantly higher than when sorption is modeled as an instantaneous process. Figure 4.9 helps to explain this observation.

In Figure 4.9, we simulate concentrations at a well slightly upgradient of the origin ($x = -0.1$ m, $y = 0$ m, $z = 0$ m) for both rate-limited ($\alpha = 0.005$ min^{-1}) and instantaneous ($\alpha = $ INF) sorption assumptions. Model 306, which was used to produce Figure 4.8 as well as Figure 4.9, assumes a Dirac pulse input at the origin at time zero. As described in Chapter 2, the Dirac function simulates an infinite concentration at the origin at $t = 0$. At very short times, concentrations are large, due to the Dirac pulse initial condition. We note from the blowup inset in the figure that concentrations are larger, and the peak arrives sooner, when sorption is rate limited. This is because at very short times (<0.5 min, which corresponds to a Damköhler number of $\frac{0.5 \text{min}}{1/0.005 \text{ min}^{-1}} = 0.0025$) the sorption process has neither significantly slowed transport nor reduced the mass in the dissolved phase; recall that the Damköhler number is the ratio of a transport timescale to a reaction timescale (in this case, the reaction is sorption), so when the Damköhler number is small, the transport processes dominate.

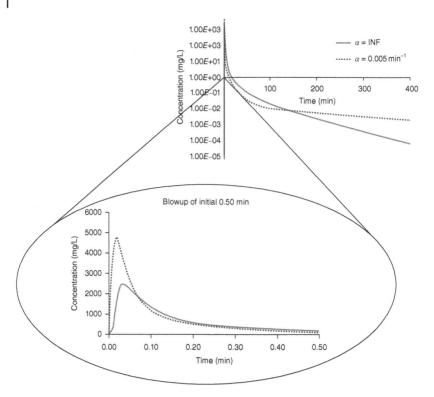

Figure 4.9 Concentration versus time output of Model 306 for Table 4.1 virtual system soil and hydraulic characteristics ($\theta = 0.25$, $\rho_b = 1.5$ kg/L, $v = 0.1$ m/min) and $M = 1$ mg, $D_x = 0.1$ m²/min, $D_y = 0.01$ m²/min, $D_z = 0.0002$ m²/min, $\lambda = 0$ min⁻¹, $k_d = 0.167$ L/kg ($R = 2$), $\alpha = $ INF versus $\alpha = 0.005$ min⁻¹. Sampling well at $x = -0.1$ m, $y = 0$ m, $z = 0$ m.

It is also interesting to note that the monitoring well at which the breakthrough curve is observed is upgradient from the initial input of chemical at the origin. Thus, the chemical is seen at the upgradient monitoring well because dispersion in the negative x-direction is larger than advection in the positive x-direction. This is not too surprising, since dispersion is modeled as a diffusive process, so mass flux is due to a concentration gradient. With a Dirac pulse, the concentration gradient is quite large, so flux in the upgradient and cross-gradient directions can be significant. Note, however, that this is a result of the model formulation, and does not reflect reality. Recall from Chapter 2 that dispersion is a construct that models spreading caused by advection (spreading due to molecular diffusion is typically ignored since it's relatively small). Thus, in reality, we would expect no chemical transport upgradient or directly cross gradient due to dispersion, even though the model simulates it. This is an example of why it's critical to understand the assumptions upon which a model is based in order to properly interpret model results!

Figure 4.9 also shows that at long times concentrations simulated at the monitoring well for the rate-limited sorption model are greater than concentrations simulated with the instantaneous sorption/desorption model. This is the result of the tailing that is caused by rate-limited desorption. Note from the apparent linearity of the semilog plot that the decrease in concentrations at long times is approximately exponential for both instantaneous and rate-limited sorption/desorption. If we calculate a first-order rate constant for the long-time decrease in concentrations, we find the rate constant for instantaneous sorption/desorption is more than three times greater than the rate constant for slow desorption. In Appendix I, the formula to calculate the long-time rate constant for the decrease in concentrations due to advection, dispersion, equilibrium sorption, and first-order degradation is derived. The long-time rate constant for rate-limited sorption is necessarily slower, as the rate-limited desorption process occurs in series with the advection, dispersion, and degradation processes, which all occur simultaneously. Note that if the rate constant for desorption is significantly slower than the rate-constant for advection, dispersion, and degradation, the desorption rate constant will be limiting. This is the case in Figure 4.9, where the desorption rate constant, α, is 0.005 min^{-1}, while the rate for advection, dispersion, and degradation, calculated using the derivation in Appendix I for $R = 1$, is 0.025 min^{-1}.

4.6 Effect of First-Order Degradation

The effect of first-order degradation on concentration versus time breakthrough curves and concentration versus distance profiles in three dimensions is similar to what we saw in Section 3.6 in one dimension. At high values of the Damköhler number, when the advection timescale (L/v or t) is considerably longer than the first-order degradation reaction timescale ($1/\lambda$), degradation dominates and concentrations are near zero. Conversely, at small Damköhler numbers, when the advection timescale is considerably shorter than the reaction timescale, advection dominates and the concentration distributions in time and space are similar to what they would be if no degradation were occurring. It is only when the advection and degradation reaction timescales are similar ($Da_I \sim 1$) that the effect of both advection and degradation on the concentration distributions in space and time may be clearly observed. These behaviors are illustrated in Figure 4.10. Figure 4.10a shows the concentration profile in space for a small Damköhler number ($Da_I = 0.02$). We observe that for this small Damköhler number, the concentration profile is relatively symmetric in the x-direction. The slight asymmetry is due to the fact that the front of the plume has been acted upon by dispersion for a longer time than the rear of the plume; we therefore see that the front exhibits slightly more spreading than the rear. We also note that the center of mass of the plume

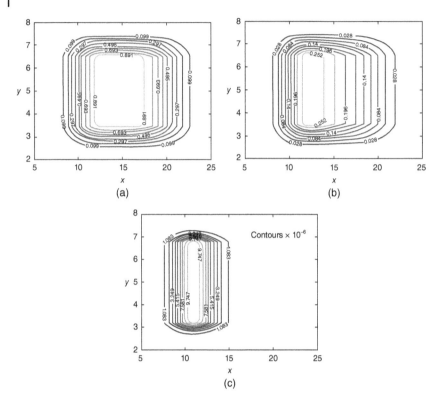

Figure 4.10 Output of Model 304 for the virtual three-dimensional system (Table 4.1) at $t = 200$ min and $t_s = 100$ min. $D_x = 0.01$ m^2/min, $D_y = 0.0001$ m^2/min, $D_z = 0.0001$ m^2/min, $k_d = 0$ L/kg, $\alpha = $ INF. (a) $\lambda = 0.0001$ min^{-1} ($Da_I = 0.02$) at $z = 3$ m, (b) $\lambda = 0.01$ min^{-1} ($Da_I = 2.0$) at $z = 3$ m, (c) $\lambda = 0.1$ min^{-1} ($Da_I = 20$.) at $z = 3$ m. All coordinates in meters and note the axes have different scales.

is approximately at $x = 15$ m. This makes sense, since a 100-min pulse time ($t_s = 100$ min) is being simulated, and the concentration versus distance snapshot is being taken at $t = 200$ min. Thus, the front of the plume has been transported by advection ($v = 0.1$ m/min) for 200 min (20 m at $v = 0.1$ m/min) and the rear of the plume has been transported with the same advective velocity for 100 min (10 m at $v = 0.1$ m/min). Thus, it makes sense that the center of the plume is found at 15 m.

Figure 4.10b, which depicts the concentration versus distance profile when the Damköhler number is on the order of 1.0 ($Da_I = 2.0$), has many important differences compared to Figure 4.10a. First, we see that the contours of Figure 4.10b are at lower concentrations than the contours of Figure 4.10a, which, of course, makes sense, since degradation is acting faster at the higher Damköhler number. We also see that the $Da_I = 2.0$ concentration distribution is quite asymmetric, compared with the relatively symmetric $Da_I = 0.02$ distribution. This is explained by the fact that compound molecules in the front of the plume have been in the system for 200 min, while compound molecules in the rear of the plume have only been in the system for 100 min. At a Damköhler number on the order of 1.0, this 100-min difference results in a significant difference in the concentration contours at the rear and front of the plume; hence, the asymmetric concentration distribution. Finally, we see that the center of mass of the plume appears to not have traveled as far as the center of mass in Figure 4.10a. Again, this is because the front of the plume has been in the system longer than the rear of the plume, degradation at the front of the plume is significantly greater than degradation at the rear, and the effect of this differential degradation is to move the center of mass in the negative x-direction.

Figure 4.10c depicts concentration versus distance contours when the Damköhler number is high ($Da_I = 20$). For high Damköhler numbers, plume concentrations are quite low (note that the contours on the plot are multiplied by a factor of 10^{-6}). Also note that at this high Damköhler number the plume is relatively symmetric, since degradation is acting so fast (relative to advection) that the difference in degradation between the front and rear of the plume is negligible. We also note that the center of the mass of the plume is at an x-value just slightly greater than 10 m. This is also a consequence of degradation being so rapid, in comparison with advection. The only molecules that remain in the system are those molecules that were released last, at the end of the 100-min pulse. Concentrations due to molecules that were released at the beginning of the pulse are negligible compared to the concentrations depicted in the figure.

Problems

4.1 Use Model 306 and the principle of superposition to simulate a concentration versus time breakthrough curve at a monitoring well located at coordinates (3, 0, 0) for Dirac pulse injections of 3 g at coordinate (0, 0, 0) and 5 g at coordinate (−1, −1, 0). Use the parameter values specified in Figure 4.7 for a sorbing, nondegrading compound. Hint: For the injection not at the origin, it will be necessary to transform the location of the monitoring well.

4.2 For the virtual three-dimensional system (Table 4.1) and the parameter values specified in Figure 4.3 for a nonsorbing, nondegrading compound, use Model 301 and the principle of superposition to simulate a concentration versus time breakthrough curve at a monitoring well located at coordinates (3, 3, 3) for injections of 3 mg/L from times 0–5 min and 10 mg/L from times 10–20 min. Hint: After running the model, it will be necessary to transform the times prior to applying superposition.

4.3 The equation in Appendix I for the long-time rate constant for the decrease in concentrations due to advection, dispersion, equilibrium sorption, and first-order degradation in a three-dimensional system (Equation (I.4)) indicates that the rate constant does not depend on dispersion in the y- and z-directions. Use Model 306 to produce semilog plots similar to Figure 4.9 to show this is true.

4.4 Derive an equation similar to Equation (I.4) for the long-time rate constant for the decrease in concentrations due to advection, dispersion, first-order degradation, and rate-limited sorption in a three-dimensional system. Keep in mind that rate-limited desorption occurs in series with the advection, dispersion, and degradation processes, which all occur simultaneously.

5

Method of Moments

In the previous chapters, we have looked at how concentration distributions vary in space and time in response to different processes, parameter values that describe these processes, and initial/boundary conditions. Another perspective that can be taken to help us think about how these concentration distributions vary in response to changing processes, parameters, and conditions is to look at the temporal and spatial moments of the distributions. Simply put, the moments of a distribution, which we define mathematically below, are quantitative measures of the shape of a distribution.

5.1 Temporal Moments

5.1.1 Definition

As we have discussed in Chapter 3, the concentration response as a function of time at a given location is known as the breakthrough curve. The absolute jth moment of a breakthrough curve (also known as the moment about the origin), which we refer to as the jth absolute temporal moment, $m_{j,t}$, is defined as

$$m_{j,t} = \int_0^\infty t^j C(x,t)dt \qquad (5.1)$$

We can immediately see from this definition that the zeroth temporal moment, $m_{0,t}$, is simply the area underneath the concentration versus time curve, with units of $[M\text{-}L^{-1}\text{-}T]$ for a one-dimensional system. It is convenient to normalize the absolute moments by the zeroth moment, so we define the jth normalized absolute temporal moment, $\mu'_{j,t}$, as follows:

$$\mu'_{j,t} = \frac{m_{j,t}}{m_{0,t}} \qquad (5.2)$$

The first normalized absolute temporal moment, $\mu'_{1,t}$, which has units of time, is the center of mass of the breakthrough curve, and it signifies the mean

Analytical Modeling of Solute Transport in Groundwater: Using Models to Understand the Effect of Natural Processes on Contaminant Fate and Transport, First Edition. Mark Goltz and Junqi Huang.
© 2017 John Wiley & Sons, Inc. Published 2017 by John Wiley & Sons, Inc.
Companion Website: www.wiley.com/go/Goltz/solute_transport_in_groundwater

residence time of molecules traveling from $x=0$ to the location where the breakthrough curve is measured. The second moment (units of time squared) is a measure of the spread of the breakthrough curve, the third a measure of the curve's skewness/asymmetry, and so on. For moments higher than the first, it is useful to define central moments (or moments about the mean) as follows:

$$\mu_{j,t} = \frac{\int_0^\infty (t - \mu'_{1,t})^j C(x,t)dt}{m_{0,t}}, \quad j \geq 2 \tag{5.3}$$

The second moment about the mean, which is also the variance of the breakthrough curve, is

$$\mu_{2,t} = \mu'_{2,t} - (\mu'_{1,t})^2 \tag{5.4}$$

5.1.2 Evaluating Temporal Moments

There are two relatively straightforward methods of analytically determining the formulas for the temporal moments of a distribution. In the first method, since we have expressions for concentration as a function of time (from Chapters 3 and 4), we can just insert those expressions into Equation (5.1) and evaluate the integral to determine the moment formulas.

A second method, which is often more convenient in that it avoids having to evaluate what could be somewhat complicated integrals, is to apply Aris' method of moments (Aris, 1958). Aris (1958) showed that

$$m_{j,t} = (-1)^j \lim_{s \to 0} \left[\frac{d^j \overline{C}(x,s)}{ds^j} \right] \tag{5.5}$$

where $\overline{C}(x,s)$ is the Laplace transform of the function $C(x,t)$, and s is the Laplace transform variable (see Appendix A for the definition of a Laplace transform of a function). Note from Appendices A and B that obtaining the Laplace time solution, $\overline{C}(x,s)$, to the Laplace transformed ADR equation and its associated transformed boundary conditions is relatively straightforward. This Laplace time solution can then be used on the right-hand side of Equation (5.5) to obtain temporal moments.

5.1.3 Temporal Moment Behavior

5.1.3.1 Advective–Dispersive Transport with First-Order Degradation and Linear Equilibrium Sorption

The solution in Laplace time for PDE (5.6) with initial/boundary conditions (5.7) is derived in Appendix J.

$$\frac{\partial C}{\partial t} = -\frac{v}{R}\frac{\partial C}{\partial x} + \frac{D_x}{R}\frac{\partial^2 C}{\partial x^2} - \frac{1}{R}\lambda C \tag{5.6}$$

$$C(x, t = 0) = 0 \tag{5.7a}$$

$$C(x = 0, t) = C_0[H(t) - H(t - t_s)] \tag{5.7b}$$

$$\frac{\partial C}{\partial x}(x = \infty, t) = 0 \tag{5.7c}$$

Equation (5.6) is the one-dimensional ADR equation assuming linear, equilibrium sorption with a retardation factor, R, and first-order degradation kinetics, with a first-order rate constant λ. Initial condition (5.7a) indicates no chemical is initially present. Boundary condition (5.7b) indicates there is a pulse of chemical at concentration C_0 at the inlet boundary ($x = 0$) from time $t = 0$ to time $t = t_s$ (see the discussion of the Heaviside step function in Section 3.2). Boundary condition (5.7c) signifies a zero concentration gradient at $x = \infty$.

From Appendix J, we see that the moment formulas for the system described by Equations (5.6) and (5.7) are

$$m_{0,t} = C_0 t_s e^{\frac{x(v-a)}{2D_x}} \tag{5.8}$$

$$\mu_{1,t} = \mu'_{1,t} = \frac{m_{1,t}}{m_{0,t}} = \frac{t_s}{2} + \frac{Rx}{a} \tag{5.9}$$

$$\mu_{2,t} = \frac{t_s^2}{12} + \frac{2D_x R^2 x}{a^3} \tag{5.10}$$

where $a = \sqrt{v^2 + 4D_x \lambda}$.

Note that the first term on the right-hand side of Equations (5.9) and (5.10) captures the effect of boundary condition (5.7b) on the first and second moments, respectively. In general (Govindaraju and Das, 2007)

$$\mu_{j,t} = \mu_{j,t}(0) + \mu_{j,t}(\delta(t)) \tag{5.11}$$

where $\mu_{j,t}(0)$ is the jth temporal moment of the boundary condition at the inlet ($x = 0$) and $\mu_{j,t}(\delta(t))$ is the jth temporal moment of the breakthrough curve at the column outlet in response to a Dirac delta pulse at the inlet (see the discussion of the Dirac delta function in Section 2.4.1).

By evaluating these formulas, we can better understand how temporal moments respond to changing parameter values. Let us define baseline conditions using the parameter values in Table 5.1 and then conduct a sensitivity analysis by varying each parameter to observe how the temporal moments respond to parameter value changes.

5.1.3.1.1 Zeroth Moment Behavior

We see from Equation (5.8) that the zeroth temporal moment, which is the area under the breakthrough curve, is a function of all parameters except R.

Figure 5.1 plots the zeroth temporal moment versus the Damköhler number ($\frac{\lambda x}{v}$). Note that at very small Damköhler numbers, when the degradation rate is

Table 5.1 Baseline parameter values for temporal moment sensitivity analysis.

Parameter	Value
Column length (x)	2 m
Chemical input concentration (C_0)	1 mg/m
Chemical pulse duration (t_s)	5 min
Retardation factor (R)	2.0
Pore water velocity (v)	0.1 m/min
Dispersivity (a_x)	0.1 m
First-order degradation rate constant (λ)	0.05 min^{-1}
Peclet number $\left(Pe = \dfrac{x}{a_x} = \dfrac{xv}{D_x} \right)$	20
Damköhler number for degradation $\left(Da_I^d = \dfrac{\lambda x}{v} \right)$	1

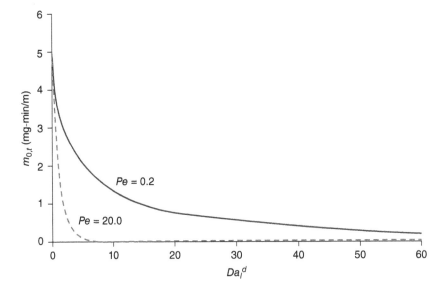

Figure 5.1 Equilibrium model: zeroth temporal moment versus the Damköhler number for degradation $\left(Da_I^d = \frac{\lambda x}{v} \right)$ for the Table 5.1 parameter values.

much less than the advection rate, the zeroth moment, with units of concentration multiplied by time, approaches the input concentration (C_0) multiplied by the input time (t_s), which for the baseline parameters is 5 mg-min/m. This is because at very small Damköhler numbers, degradation is negligible, so the area of the breakthrough curve at the system outlet equals the area of the input to the system at $x = 0$. At very large Damköhler numbers, degradation is much faster than advection, so concentrations are negligible by the time the chemical reaches the system outlet, and the zeroth moment approaches 0. The effect of spreading (quantified by the Peclet number, Pe) on the zeroth moment is also shown in Figure 5.1. At low Pe, where spreading is significant, the zeroth moment decreases less than it does at high Pe. This is because with less spreading (high Pe), overall concentrations as the chemical moves through the system are higher, and, therefore, since the change in concentration is proportional to concentration for a first-order reaction, more mass is lost in the case where spreading is small, and the zeroth moment decreases more. This is explicitly shown in Figure 5.2, where the zeroth moment decreases as Pe increases. Of course, in the case where there is no degradation (as the Damköhler number approaches zero), there is no effect of dispersion on the zeroth moment and the zeroth moment is constant at $C_0 * t_s$ independent of Pe.

5.1.3.1.2 First Moment Behavior

From Figure 5.3, we see that independent of Pe, when there is no degradation ($Da_I = 0$), the first moment is constant at 42.5 min (per Equation (5.9)). As the Da_I increases, the first moment decreases. That is, the faster the degradation, the shorter the mean residence time of molecules in the column is. This can be explained if we think about how degradation affects molecules while they are in the column. Due to dispersion, some molecules will be in the column

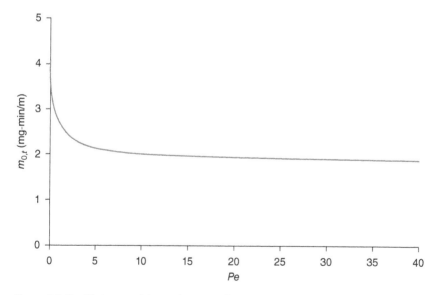

Figure 5.2 Equilibrium model: zeroth temporal moment versus the Peclet number $\left(Pe = \frac{vx}{D_x}\right)$ for the Table 5.1 parameter values.

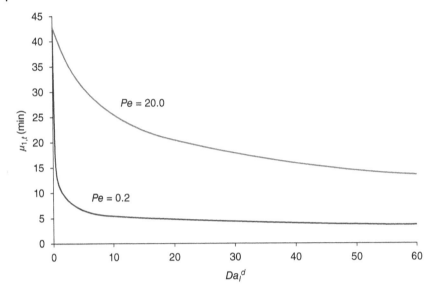

Figure 5.3 Equilibrium model: first temporal moment versus the Damköhler number for degradation $\left(Da_l^d = \frac{\lambda x}{v}\right)$ for the Table 5.1 parameter values.

longer than others. The molecules that are in the column longer (represented by the tailing portion of the breakthrough curve) will degrade more than molecules that are in the column for shorter periods (the leading portion of the breakthrough curve). Thus, as the degradation rate increases, degradation of the tailing portion of the breakthrough curve increases and the center of mass of the curve moves to earlier times (i.e., the first moment of the curve decreases). Figure 5.4 is consistent with this explanation. Figure 5.4 shows that the more spreading there is, the more the first moment decreases. When there is no spreading ($Pe \to \infty$), the first moment is the same as when there is no degradation, since all chemical molecules remain in the column equal times, and all are subject to the same amount of degradation (so the first moment, which is just the center of gravity of the breakthrough curve, is unaffected by the degradation).

Figure 5.5 demonstrates that, as is obvious from Equation (5.9), the first moment increases proportionally with the retardation factor. The higher the value of R, the slower the chemical moves through the system and the higher the first moment.

5.1.3.1.3 Second Moment Behavior
Before discussing how the second temporal moment behaves as a function of transport parameters, let us first nondimensionalize Equation (5.10) by dividing

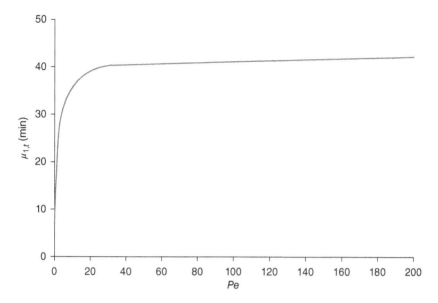

Figure 5.4 Equilibrium model: first temporal moment versus the Peclet number $\left(Pe = \frac{vx}{D_x}\right)$ for the Table 5.1 parameter values.

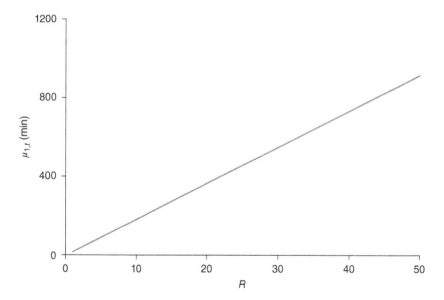

Figure 5.5 Equilibrium model: first temporal moment versus the retardation factor (R) for the Table 5.1 parameter values.

all the terms in the equation by the square of the characteristic timescale (x/v):

$$\overline{\mu}_{2,t} = \frac{t_s^2}{12}\frac{v^2}{x^2} + \frac{2R^2\sqrt{Pe}}{(Pe + 4Da_I^d)^{3/2}} \tag{5.12}$$

where

$\overline{\mu}_{2,t}$ = dimensionless second temporal moment about the mean = $\mu_{2,t}\frac{v^2}{x^2}$

Pe = Peclet number = $\frac{vx}{D_x}$

Da_I^d = Damköhler number for degradation = $\frac{\lambda x}{v}$

Figure 5.6 shows that when we can ignore degradation (low Da_I^d), spreading is controlled by the Pe. Obviously, as Pe decreases, spreading (as indicated by the second temporal moment) increases. As degradation becomes more important (the Da_I^d increases), spreading decreases, because the tailing portions of the breakthrough curves are degraded. Interestingly, at a given high value of Da_I^d, degradation has a bigger effect on the second temporal moment of the low Pe breakthrough curve. Mathematically, we can see this by looking at the second term on the right-hand side of Equation (5.10) or (5.12). For given relative values of the advection and degradation rates (i.e., Da_I^d is specified), as D_x, the dispersion coefficient, increases (i.e., Pe decreases), the second

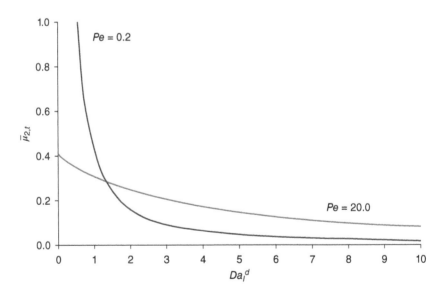

Figure 5.6 Equilibrium model: dimensionless second temporal moment versus the Damköhler number for degradation $\left(Da_I^d = \frac{\lambda x}{v}\right)$ for the Table 5.1 parameter values at two different values of Pe.

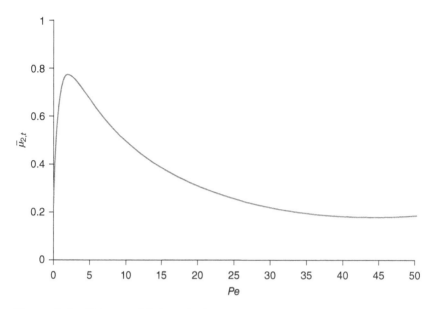

Figure 5.7 Equilibrium model: dimensionless second temporal moment versus the Peclet number $\left(Pe = \frac{vx}{D_x}\right)$ for the Table 5.1 parameter values.

moment decreases. This can be understood when we realize that as dispersion increases, the "opportunity" for degradation to act on the contaminant in the breakthrough curve tail also increases, thereby resulting in a smaller second temporal moment. Another way to think about this is to consider what would happen if there were no dispersion ($Pe \to \infty$). In that case, increasing the degradation would have no effect on the second temporal moment – as can be seen immediately by looking at Equation (5.10) as $D_x \to 0$ or Equation (5.12) as $Pe \to \infty$.

Figure 5.7 illustrates some interesting behavior of the second temporal moment as a function of Pe. We realize that in the absence of degradation, as dispersion increases (Pe decreases), the second temporal moment should increase. However, as discussed above, degradation acting in concert with dispersion results in a decrease of the second temporal moment, due to degradation of contaminant in the breakthrough curve tail. Thus, for given advection, dispersion, and degradation parameter values, there will be some value of dispersion (quantified by Pe) where spreading (quantified by the second temporal moment) is a maximum. We see this behavior in Figure 5.7. At high Pe, spreading increases as Pe decreases. However, eventually, there comes a point where the increase in spreading due to the decrease in Pe is counteracted by the decrease in spreading caused by the combination of degradation and dispersion (quantified by the $4D_x\lambda$ term in the denominator

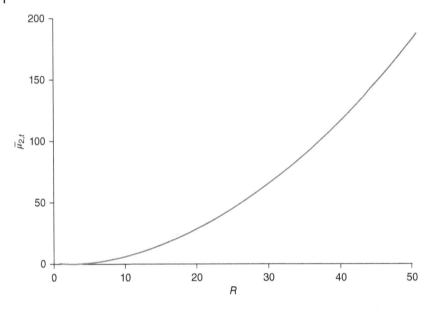

Figure 5.8 Equilibrium model: dimensionless second temporal moment versus the retardation factor (R) for the Table 5.1 parameter values.

of Equation (5.10)). At this point, spreading, measured by the second temporal moment, is a maximum. As Pe continues to decrease, spreading decreases as well. In Equations (5.10) and (5.12), by setting $\frac{d\mu_{2,t}}{dPe} = \frac{d\bar{\mu}_{2,t}}{dPe} = 0$, we see this point of maximum spreading occurs when $Pe = 2Da_1^d$.

For the parameter values used to construct Figure 5.7, the maximum second moment occurs when $Pe = 2$.

From Equations (5.10) and (5.12), it may immediately be seen that the second temporal moment increases with the square of R. This relation is demonstrated in Figure 5.8. The reason spreading increases with R is obviously due to the fact that the longer the contaminant is in the column, the more opportunity there is for spreading. The reason the increase in second moment is a function of R^2 is less apparent. To better understand this, let us rewrite the second term on the right-hand side of Equation (5.10), assuming no degradation.

$$\frac{2D_x R^2 x}{v^3} = \frac{2\left(\dfrac{D_x}{R}\right)\left(R\dfrac{x}{v}\right)}{\left(\dfrac{v}{R}\right)^2} \tag{5.13}$$

Note that the second term on the right-hand side of Equation (5.10) represents the contribution to the second temporal moment of the combination of advection, dispersion, and sorption. It is well known that the second temporal moment for a nonsorbing, conservative tracer is related to the pore velocity (v)

and dispersion coefficient (D_x) as follows (e.g., Domenico and Schwartz, 1998):

$$\mu_{2,t} = \frac{2D_x t}{v^2} = \frac{2D_x \left(\frac{x}{v}\right)}{v^2} \tag{5.14}$$

Equation (5.13) is the same as Equation (5.14), but (5.13) is for a sorbing solute, where the pore velocity (v) and dispersion coefficient (D_x) of a non-sorbing tracer are divided by the sorbing solute retardation factor (R), and the residence time of a nonsorbing tracer in the column $(t = \frac{x}{v})$ is multiplied by R, hence, the relationship between second temporal moment and R^2.

5.1.3.2 Advective–Dispersive Transport with First-Order Degradation and Rate-Limited Sorption

As shown in Section 2.3.2.2.1, the coupled partial differential equations (5.15a) and (5.15b) describe advective–dispersive transport with first-order degradation and rate-limited sorption. The four terms on the right-hand side of Equation (5.15a) model advection, dispersion, first-order decay, and rate-limited sorption, respectively. Equation (5.15b) models the sorption reaction as a first-order process with rate constant α.

$$\frac{\partial C}{\partial t} = -v\frac{\partial C}{\partial x} + D_x \frac{\partial^2 C}{\partial x^2} - \lambda C - \frac{\rho_b}{\theta}\frac{\partial S}{\partial t} \tag{5.15a}$$

$$\frac{\partial S}{\partial t} = \alpha(k_d C - S) \tag{5.15b}$$

$$C(x, t = 0) = S(x, t = 0) = 0 \tag{5.16a}$$

$$C(x = 0, t) = C_0[H(t) - H(t - t_s)] \tag{5.16b}$$

$$\frac{\partial C}{\partial x}(x = \infty, t) = 0 \tag{5.16c}$$

For initial condition (5.16a) and boundary conditions (5.16b) and (5.16c), Equation (5.15) can be solved in Laplace time, and Aris' method of moments applied (see Appendix J), to obtain formulas for the zeroth, first, and second temporal moments of the breakthrough curves (Equations (5.17), (5.18), and (5.19)), respectively.

$$m_{0,t} = C_0 t_s e^{\frac{x(v-a)}{2D_x}} \tag{5.17}$$

$$\mu'_{1,t} = \frac{m_{1,t}}{m_{0,t}} = \frac{t_s}{2} + \frac{Rx}{a} \tag{5.18}$$

$$\mu_{2,t} = \frac{t_s^2}{12} + \frac{2D_x R^2 x}{a^3} + \frac{2(R-1)x}{\alpha a} \tag{5.19}$$

where $a = \sqrt{v^2 + 4D_x \lambda}$ and $R = 1 + \frac{\rho_b k_d}{\theta}$.

5.1.3.2.1 Zeroth and First Moment Behavior

As may be seen by comparing Equations (5.8) and (5.9) with Equations (5.17) and (5.18), the zeroth and first temporal moments of the breakthrough curves are the same, regardless of whether sorption is modeled as an equilibrium or rate-limited process. Thus, the previous discussions in Sections 5.1.3.1.1 and 5.1.3.1.2 regarding the zeroth and first temporal moment behavior when equilibrium sorption is assumed also apply when sorption is rate limited. We also see from Equations (5.17) and (5.18) that the zeroth and first temporal moments are independent of the rate constant, α. This is perhaps a nonintuitive result. Figure 3.10 shows breakthrough curves at three different values of α. As noted in Chapter 3, although not apparent, the area under all three breakthrough curves in Figure 3.10 is the same. This must be the case, as a result of Equation (5.17), which shows the zeroth moment is not a function of the rate constant. What is also true is that the center of mass (the first moment) of all three breakthrough curves is the same. By using Equation (5.18) with the parameter values for Figure 3.10, we calculate the first temporal moment for all three breakthrough curves to be 42.5 min. Looking at the curves, and especially the curve for $\alpha \rightarrow 0$, this does not appear correct. However, the first moment for all three curves is indeed 42.5 min. This is a result of breakthrough curve tailing. Tailing is most apparent in the $\alpha = 0.05\,\text{min}^{-1}$ curve. However, even the $\alpha \rightarrow 0$ curve has tailing, though it is not obvious from the figure. At very slow sorption/desorption rates, only a small fraction of chemical sorbs. However, once sorbed, the timeframe for desorption is quite long, resulting in the (imperceptible) breakthrough curve tail. This tailing leads to the first moment being much larger than is apparent from visual inspection of the breakthrough curve.

5.1.3.2.2 Second Moment Behavior

Comparing Equation (5.10) for equilibrium sorption and Equation (5.19) for rate-limited sorption, we see that the second temporal moment calculated for the rate-limited case is larger, due to the extra spreading of the breakthrough curve that results from the rate-limited sorption (quantified by the third term on the right-hand side of Equation (5.19)).

As before, let us nondimensionalize Equation (5.19) by dividing all the terms in the equation by the square of the characteristic timescale (x/v):

$$\bar{\mu}_{2,t} = \frac{t_s^2\,v^2}{12\,x^2} + \frac{2R^2\sqrt{Pe}}{(Pe + 4Da_I^d)^{3/2}} + \frac{2(R-1)\sqrt{Pe}}{Da_I^s\sqrt{Pe + 4Da_I^d}} \tag{5.20}$$

where

$\bar{\mu}_{2,t}$ = dimensionless second temporal moment about the mean = $\mu_{2,t}\dfrac{v^2}{x^2}$

Pe = Peclet number = $\dfrac{vx}{D_x}$

Da_I^d = Damköhler number for degradation = $\frac{\lambda x}{v}$

Da_I^s = Damköhler number for sorption = $\frac{\alpha x}{v}$

R = retardation factor = $1 + \frac{\rho_b k_d}{\theta}$

Note that because there are two rate constants in the equation (the degradation rate constant, λ, and the sorption rate constant, α) there are two Damköhler numbers. The Damköhler number for degradation (Da_I^d) is the ratio of the advection timescale (x/v) and the degradation timescale (λ^{-1}), while the Damköhler number for sorption (Da_I^s) is the ratio of the advection timescale and the sorption timescale (α^{-1}).

Figure 5.9 shows the relation between spreading, as quantified by the dimensionless second temporal moment about the mean, $\overline{\mu}_{2,t}$, and the retardation factor, R. We see from the figure and Equation (5.20) that spreading increases quadratically as a function of R. The relation with R^2 was discussed in Section 5.1.3.1.3 for equilibrium sorption. In the case of rate-limited sorption, a term is added, which is proportional to R, to account for the additional spreading resulting from sorption kinetics.

Figure 5.10 shows the relationship between spreading, as quantified by the dimensionless second temporal moment about the mean, $\overline{\mu}_{2,t}$, and the Damköhler number for sorption, Da_I^s, for two values of Pe. At very low values of Da_I^s spreading is large, as the third term on the right-hand side of Equation (5.20) grows large with decreasing Da_I^s. Note that Da_I^s approaches

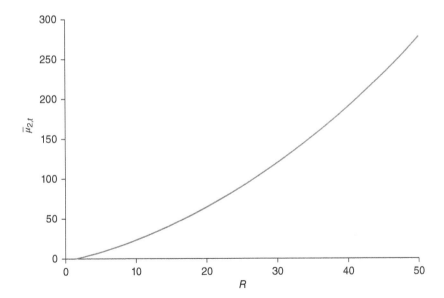

Figure 5.9 Rate-limited sorption model: second temporal moment versus the retardation factor (R) for the Table 5.1 parameter values with $\alpha = 0.05$ min^{-1} ($Da_I^s = \alpha x/v = 1$).

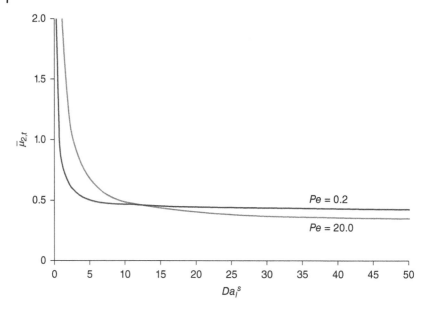

Figure 5.10 Rate-limited sorption model: dimensionless second temporal moment versus the Damköhler number for sorption ($Da_I^s = \alpha x/v$) for the Table 5.1 parameter values at two different values of *Pe*.

zero as $\alpha \to 0$, so the case of small Da_I^s corresponds to the scenario where sorption is extremely slow. It is the imperceptible breakthrough curve tail discussed in Section 5.1.3.2.1 and (not) seen in the Figure 3.10 $\alpha \to 0$ breakthrough curve, which results in very large values of the second temporal moment.

As the rate of sorption, and therefore, Da_I^s, increases, the effect of the third term on the right-hand side of Equation (5.20) decreases, and the second temporal moment decreases. Eventually, as $Da_I^s \to \infty$, sorption is described by an equilibrium isotherm, and the second moment calculated using the rate-limited sorption model (Equation (5.20)) is the same as the second moment calculated using the equilibrium model (Equation (5.12)).

Comparing the two curves in Figure 5.10, we see that as $Da_I^s \to \infty$ and spreading due to sorption kinetics is unimportant, the curve with the lower *Pe* (higher dispersion) has a larger second moment, as would be expected, since spreading is due only to dispersion when sorption is fast. However, when $Da_I^s \to 0$ and spreading due to sorption kinetics dominates, the curve with the higher *Pe* (lower dispersion) has a larger second moment. The physical reason for this is that lower dispersion results in higher dissolved chemical concentrations, which drives more chemical into the sorbed phase, thereby increasing the impact of the tail of the breakthrough curve on the second moment as the chemical desorbs.

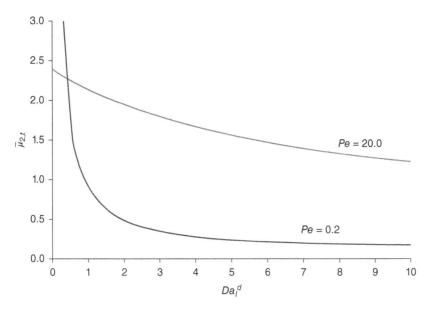

Figure 5.11 Rate-limited sorption model: dimensionless second temporal moment versus the Damköhler number for degradation $\left(Da_I^d = \frac{\lambda x}{v} \right)$ for the Table 5.1 parameter values with $\alpha = 0.05 \ \text{min}^{-1}$ $(Da_I^s = \alpha x/v = 1)$ at two different values of Pe.

Figure 5.11 shows that when we can ignore degradation (low Da_I^d), spreading is controlled by Pe and sorption kinetics. Looking at Equation (5.20), we see that as $Da_I^d \to 0$, the second term on the right-hand side is inversely proportional to Pe and the third term on the right-hand side is no longer a function of Pe and is only a function of R and Da_I^s. Obviously, as Pe decreases, spreading (as indicated by the second temporal moment) increases. As degradation becomes more important (the Da_I^d increases), spreading decreases, because the tailing portions of the breakthrough curves are degraded. For the same reasons as discussed in Section 5.1.3.1.3 for equilibrium sorption, at a given high value of Da_I^d, degradation has a bigger effect on reducing the second temporal moment of the low Pe breakthrough curve compared with the high Pe curve.

For rate-limited sorption, Figure 5.12 exhibits the same interesting behavior of the second temporal moment as a function of Pe that we saw earlier for equilibrium sorption (Figure 5.7). As we found for equilibrium sorption, there is a value of Pe at which the second temporal moment is a maximum. At Pe below this value, degradation acting in concert with dispersion results in a decrease of the second temporal moment, due to degradation of contaminant in the breakthrough curve tail. At Pe above this value, the second temporal moment decreases as Pe increases. Using Equation (5.20), and setting $\frac{d\bar{\mu}_{2,t}}{dPe} = 0$,

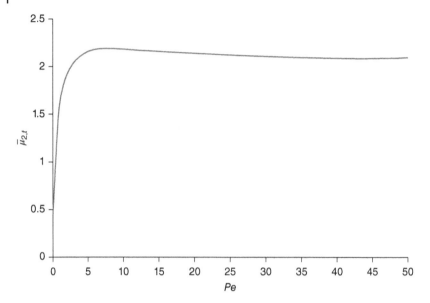

Figure 5.12 Rate-limited sorption model: dimensionless second temporal moment versus the Peclet number $\left(Pe = \frac{vx}{D_x} \right)$ for the Table 5.1 parameter values with $\alpha = 0.05$ min^{-1} $(Da_I^s = \alpha x / v = 1)$.

we see this point of maximum spreading occurs when

$$Pe = \frac{2Da_I^d Da_I^s R^2 + 8(R-1)(Da_I^d)^2}{Da_I^s R^2 - 2(R-1)Da_I^d} \tag{5.21}$$

For the parameter values used to construct Figure 5.12, the maximum second moment occurs when $Pe = 8$.

5.2 Spatial Moments

This spatial moment discussion will parallel our previous discussion of temporal moments.

5.2.1 Definition

The absolute jth moment of a concentration profile in space (also known as the moment about the origin), which we refer to as the jth absolute spatial moment of the concentration distribution, m_j, is defined as

$$m_j = \int_{-\infty}^{\infty} x^j C(x, t) dx \tag{5.22}$$

We can see from this definition that the zeroth spatial moment, m_0, is simply the area underneath the concentration versus x-coordinate curve, with units of mass [M] for a one-dimensional system. It is convenient to normalize the absolute moments by the zeroth moment, so we define the jth normalized absolute spatial moment, μ'_j, as follows:

$$\mu'_j = \frac{m_j}{m_0} \tag{5.23}$$

The first normalized absolute spatial moment, μ'_1, which has units of length [L], is the location of the center of mass of the concentration distribution curve, and it signifies the average distance traveled by solute molecules over time, t. The second moment (units of length squared) is a measure of the spread of the concentration distribution, the third a measure of the curve's skewness/asymmetry, and so on. For moments higher than the first, it is useful to define central moments (or moments about the mean) as follows:

$$\mu_j = \frac{\int_{-\infty}^{\infty} (x - \mu'_1)^j C(x, t)dx}{m_0}, \quad j \geq 2 \tag{5.24}$$

The second moment about the mean, which is also the variance of the concentration distribution, is

$$\mu_2 = \mu'_2 - (\mu'_1)^2 \tag{5.25}$$

5.2.2 Evaluating Spatial Moments

As with temporal moments, there are two relatively straightforward methods of analytically determining the formulas for the spatial moments of a distribution. In the first method, since we have expressions for concentration as a function of space (from Chapters 3 and 4), we can just insert those expressions into Equation (5.22) and evaluate the integral to determine the moment formulas.

A second method, which is often more convenient in that it avoids having to evaluate what could be somewhat complicated integrals, is to apply a modified version of Aris' method of moments (Goltz and Roberts, 1987):

$$m_j = i^j \lim_{p \to 0} \left[\frac{d^j \overline{F}(p, t)}{dp^j} \right] \tag{5.26}$$

$\overline{F}(p, t)$ is the Fourier transform of the function $C(x, t)$, defined as

$$\overline{F}(p.t) = \int_{-\infty}^{\infty} e^{-ipx} C(x, t)dx \tag{5.27}$$

where p is the Fourier transform variable and i is the imaginary unit. Goltz and Roberts (1987) derived the Fourier domain solution to the ADR equation and applied Equation (5.26), along with Equations (5.23) and (5.24) to obtain spatial moment formulas.

5.2.3 Spatial Moment Behavior

5.2.3.1 Advective–Dispersive Transport with First-Order Degradation and Linear Equilibrium Sorption

The solution in Fourier space for PDE (5.28) is derived in Appendix K.

$$\frac{\partial C}{\partial t} = -\frac{v}{R}\frac{\partial C}{\partial x} + \frac{D_x}{R}\frac{\partial^2 C}{\partial x^2} - \frac{1}{R}\lambda C \tag{5.28}$$

$$C(x, t = 0) = \frac{M}{R}\delta(x = 0) \tag{5.28a}$$

$$\frac{\partial C}{\partial x}(x = \pm\infty, t) = 0 \tag{5.28b}$$

Equation (5.28) is the one-dimensional ADR equation assuming linear, equilibrium sorption with a retardation factor, R, and first-order degradation kinetics, with a first-order rate constant λ. Initial condition (5.28a) indicates at time zero there is a Dirac pulse of total mass M and dissolved mass M/R at the origin of an infinite system. Boundary condition (5.28b) indicates zero concentration gradients at $x = \pm\infty$.

From Appendix K, we see that the spatial moments for the conditions modeled by Equation (5.28) are

$$m_0 = \frac{M}{R}e^{-\frac{\lambda}{R}t} \tag{5.29}$$

$$\mu_1' = \frac{m_1}{m_0} = \frac{v}{R}t \tag{5.30}$$

$$\mu_2 = \mu_2' - (\mu_1')^2 = \frac{m_2}{m_0} - \left(\frac{m_1}{m_0}\right)^2 = 2\frac{D_x}{R}t \tag{5.31}$$

Unlike temporal moment behavior, where each moment of the concentration breakthrough curve depends on a combination of the advection (v), dispersion (D_x), and degradation (λ) parameters (see Equations (5.8)–(5.10)), interpretation of spatial moment behavior is more straightforward. The zeroth spatial moment strictly depends on the degradation parameter, which is modified by the retardation factor since degradation is assumed to take place in the dissolved phase only. The first normalized spatial moment depends on the advection parameter, again modified by the retardation factor to account for slowing of the plume due to equilibrium sorption to an immobile phase. Finally, we see the second normalized spatial moment about the mean is only a function of the retardation factor–modified dispersion parameter, where again, spreading of the plume (as quantified by the second moment) is slowed due to equilibrium sorption to an immobile phase.

As is apparent from Equation (5.29), a plot of the zeroth spatial moment versus time exhibits exponential decay, with a first-order rate constant equal to λ/R. Equations (5.30) and (5.31) show that plots of the first and second moments versus time are linear, with slopes of v/R and $2D_x/R$, respectively.

5.2.3.2 Advective–Dispersive Transport with First-Order Degradation and Rate-Limited Sorption

The solution in Fourier space for PDE (5.32) is derived in Appendix L.

$$\frac{\partial C}{\partial t} = -v\frac{\partial C}{\partial x} + D_x\frac{\partial^2 C}{\partial x^2} - \lambda C - \frac{\rho_b}{\theta}\frac{\partial S}{\partial t} \tag{5.32}$$

$$\frac{\partial S}{\partial t} = \alpha(k_d C - S) \tag{5.32a}$$

$$C(x, t = 0) = M\delta(x) \tag{5.32b}$$

$$S(x, t = 0) = 0 \tag{5.32c}$$

$$\frac{\partial C}{\partial x}(x = \pm\infty, t) = 0 \tag{5.32d}$$

Equation (5.32) is the one-dimensional ADR equation with first-order degradation (rate constant λ) and rate-limited sorption (first-order rate constant α). Initial condition (5.32b) indicates at time 0 there is a Dirac pulse of mass M input into the dissolved phase at the origin of an infinite system, initial condition (5.32c) indicates at time 0 there is no sorbed phase compound, and boundary condition (5.32d) indicates zero concentration gradients at $x = \pm\infty$. Equation (L.2) is the solution in Fourier space.

From Appendix L, we see that the spatial moments for the conditions modeled by Equation (5.32) are

$$m_0 = M[\Delta_1 e^{s_{10}t} - \Delta_2 e^{s_{20}t}] \tag{5.33}$$

$$m_1 = -M(\Omega_1 e^{s_{10}t} - \Omega_2 e^{s_{20}t}) \tag{5.34}$$

$$m_2 = -M(\Psi_1 e^{s_{10}t} - \Psi_2 e^{s_{20}t}) \tag{5.35}$$

where the parameters in Equations (5.33)–(5.35) are defined in Appendix L. We then apply Equations (5.23) and (5.25) to the absolute moment formulas to obtain the mean (μ_1') and variance (μ_2) of the concentration distributions in space simulated by model (5.32).

In the following sections, we illustrate the behavior of the spatial moments over time by plotting simulations of the zeroth spatial moment, as well as the mean and variance of the spatial concentration distributions. Baseline values of the parameters used for the simulations are given in Table 5.2.

Table 5.2 Baseline parameter values for spatial moment sensitivity analysis.

Parameter	Value
Chemical input mass (M)	100 mg
Porosity (θ)	0.3
Sorption distribution coefficient (k_d)	0.3 L/kg
Porous media bulk density (ρ_b)	1.76 kg/L
Pore water velocity (v)	2.0 m/d
Dispersivity (a_x)	0.25 m
First-order mass transfer rate constant (α)	0.20 d^{-1}
First-order degradation rate constant (λ)	0.20 d^{-1}

5.2.3.2.1 Spatial Moment Behavior When There Is Rate-Limited Sorption But No Degradation

Zeroth Spatial Moment Behavior When There Is Rate-Limited Sorption But No Degradation Figure 5.13 shows the behavior of the zeroth spatial moment over time when there is no degradation. We see from the figure that the zeroth moment decreases, even though there is no degradation. This is due to the fact that we are plotting the zeroth moment **of the dissolved phase** over time, and the moment is reduced as mass is transferred from the dissolved to the sorbed phase. We note that for the simulations where $\alpha \neq 0$, the steady-state mass in the dissolved phase (m_0^∞) is equal to the initial mass divided by the retardation factor. That is, $m_0^\infty = \frac{M}{1+\beta} = \frac{M}{R}$ where $\beta = \frac{\rho_b k_d}{\theta}$. For Table 5.2 parameter values, $m_0^\infty = \frac{100\,\text{mg}}{2.76} = 36.2$ mg. This is because at long times ($Da_I^s \gg 1$) sorption may be viewed as an equilibrium process, so the ratio of total mass to mass in the dissolved phase is quantified by the retardation factor.

When $\alpha = 0$, we see that there is no mass loss over time, since without either degradation or mass transfer to the sorbed phase, there is no loss of mass in the dissolved phase.

First Spatial Moment Behavior When There Is Rate-Limited Sorption But No Degradation Figure 5.14 shows the behavior of the first normalized spatial moment (μ_1') over time when there is no degradation ($\lambda = 0$) for Table 5.2 parameter values. Note the slope of the curves in Figure 5.14 indicates the velocity of the dissolved plume, and we see that except for the curve with no sorption ($\alpha = 0$) the velocity is decreasing over time, eventually approaching a constant velocity at large time ($Da_I^s \gg 1$). The long-time plume velocity is the same as the velocity of a plume where sorption is assumed to be instantaneous (i.e., equilibrium sorption may be assumed). That is, even though sorption

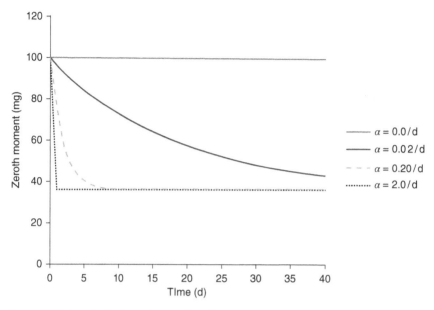

Figure 5.13 Rate-limited sorption model with no degradation ($\lambda = 0.0$): zeroth spatial moment versus time for the Table 5.2 parameter values with $\alpha = 0.0, 0.02, 0.20$, and $2.0\,d^{-1}$.

is rate limited, at long times equilibrium sorption may be assumed and the long-time plume velocity is the groundwater pore velocity divided by the retardation factor. Thus, we see that except for the curve with no sorption ($\alpha = 0$) the slope of the first spatial moment versus time plots shown in Figure 5.14a are $v/(1 + \beta) = v/\left(1 + \frac{\rho_b k_d}{\theta}\right) = v/R$ at long times ($t \gg 1/\alpha$ or $Da_I^s \gg 1$). From Figure 5.14b, we see that at short times ($t \ll 1/\alpha$ or $Da_I^s \ll 1$), the plume velocity is the groundwater pore velocity, unretarded by sorption. This plume deceleration behavior was observed in a field study of contaminant transport and was attributed to rate-limited sorption (Goltz and Roberts, 1987).

Second Spatial Moment Behavior When There Is Rate-Limited Sorption But No Degradation Figure 5.15 shows the behavior of the second spatial moment about the mean (μ_2) over time for Table 5.2 parameter values when there is no degradation. Note from Equation (5.31) that the slope of the curves in Figure 5.15 indicates twice the effective dispersion of the plume. We see that except for the curve with no sorption ($\alpha = 0$) the slope (and therefore the dispersion) is changing over time, eventually approaching a constant at large time ($t \gg 1/\alpha$ or $Da_I^s \gg 1$). Goltz and Roberts (1987) showed that at small times ($t \ll 1/\alpha$ or $Da_I^s \ll 1$) for a nondegrading compound affected by rate-limited sorption, the effective dispersion (D_{eff}) of the plume was described by

$$D_{eff} = D_x \tag{5.36}$$

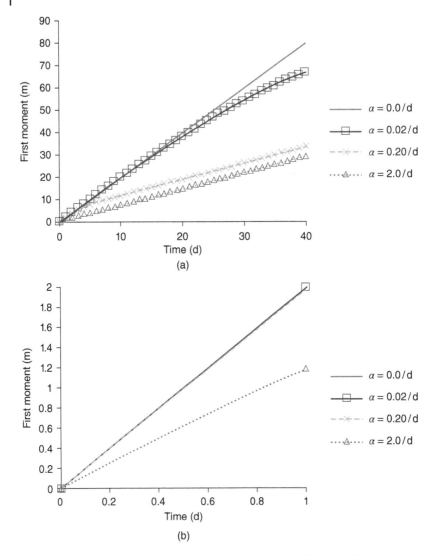

Figure 5.14 Rate-limited sorption model with no degradation ($\lambda = 0.0$): (a) first spatial moment versus time for the Table 5.2 parameter values with $\alpha = 0.0, 0.02, 0.20,$ and $2.0\,\mathrm{d^{-1}}$, (b) detail showing behavior at small times.

and at large times ($t \gg 1/\alpha$ or $Da_l^s \gg 1$)

$$D_{\text{eff}} = \frac{D_x}{1 + \beta} + \frac{v^2\beta}{\alpha(1 + \beta)^3} \tag{5.37}$$

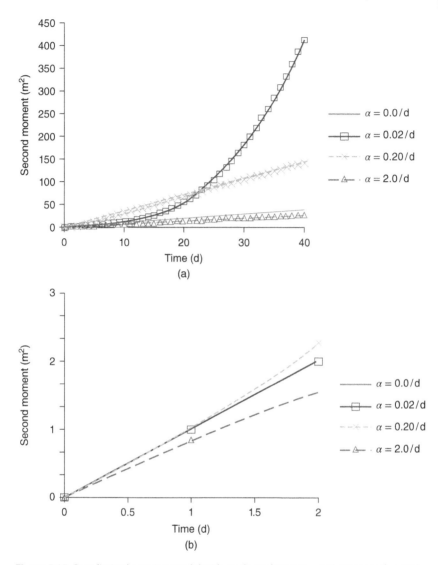

Figure 5.15 Rate-limited sorption model with no degradation ($\lambda = 0.0$): (a) second spatial moment versus time for the Table 5.2 parameter values with $\alpha = 0.0, 0.02, 0.20,$ and $2.0\,d^{-1}$, (b) detail showing behavior at small times.

Note that depending on the relative values of the parameters, the effective dispersion may increase or decrease over time. This is seen in Figure 5.15. From Figure 5.15b, we see that at small times, for all values of the mass transfer rate constant, $D_{\text{eff}} = D_x = 0.5 \ \text{m}^2/d$, which is half of the slope. At large times, the effective dispersion may be determined by calculating half the long-time slope, and it is found to be 0.5, 1.7, 1.9, and 0.35 m^2/d for $\alpha = 0.0, 0.02, 0.2$, and $2.0\,d^{-1}$, respectively. Except for the no sorption case, this is consistent with the value of D_{eff} that is calculated using Equation (5.37). For the nonsorbing compound ($\alpha = 0$), D_{eff} is constant and Equation (5.37) does not apply (since Equation (5.37) is derived for a sorbing, nondegrading compound).

5.2.3.2.2 Spatial Moment Behavior When There Is Rate-Limited Sorption and Degradation

Zeroth Spatial Moment Behavior When There Is Rate-Limited Sorption and Degradation Since α, λ, and β are all positive numbers, we can see from Equation (L.3) that s_{10} and s_{20} in Equation (5.33) are both negative, and $|s_{20}| > |s_{10}|$. Therefore, Equation (5.33) tells us that the zeroth spatial moment is the difference between two exponentially decaying terms whose magnitudes depend on the relative values of the degradation (λ) and mass transfer (α) rate constants, as well as the affinity of the compound for the immobile (sorbed) phase (β). For a given α it is evident that as λ increases, the faster the exponential decay of the zeroth moment is. This is seen in Figure 5.16. Of course, when $\lambda = 0$ we see the same result that we saw previously in Figure 5.13 for our simulations of rate-limited sorption with no decay; the steady-state mass in the dissolved phase (m_0^∞) is equal to the initial mass divided by the retardation factor.

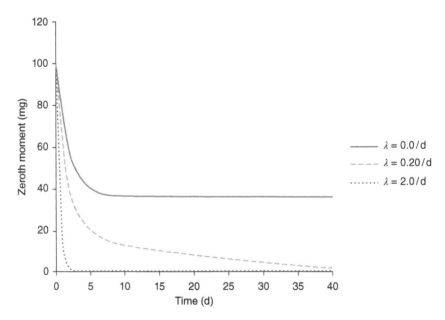

Figure 5.16 Rate-limited sorption model with degradation. Zeroth spatial moment versus time for the Table 5.2 parameter values with $\lambda = 0.0, 0.20$, and $2.0\,d^{-1}$.

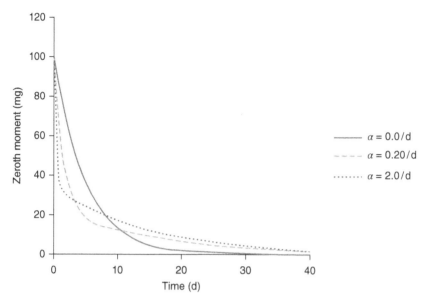

Figure 5.17 Rate-limited sorption model with degradation. Zeroth spatial moment versus time for the Table 5.2 parameter values with $\alpha = 0.0$, 0.20, and $2.0\,\mathrm{d^{-1}}$.

The relationship between increasing α for a given value of λ is somewhat more complex. Because degradation is assumed to occur in the dissolved phase only, and the zeroth moment is the zeroth moment of the dissolved phase, the mass transfer rate constant affects the zeroth moment in two, opposing, ways. As the rate constant increases, mass moves more quickly out of the mobile (dissolved) phase and, therefore, the zeroth moment decreases more rapidly. On the other hand, though, since mass in the immobile (sorbed) phase is not degraded, more mass is "preserved" in the system, and as this mass desorbs into the mobile phase, the rate of decrease of the zeroth moment slows. These effects are exhibited in Figure 5.17. We observe that at short times, the zeroth moment decreases more rapidly as the mass transfer rate constant increases. However, at long times, an increased mass transfer rate constant results in a larger "tail" for the zeroth moment versus time plot.

The effect of increasing β on the shape of the zeroth moment versus time plot is seen in Figure 5.18 to be similar to the effect of increasing α. For a given value of α, at early times, more mass moves more quickly into the immobile zone as β increases. This is a consequence of Equation (5.32a) that describes mass transfer between the mobile and immobile phases. At early times, when the value of the immobile phase concentration, S, is low, the approximate rate at which mass moves from the mobile to immobile phase ($\partial S/\partial t$) is the dissolved phase concentration, C, multiplied by the product of the mass transfer rate constant, α,

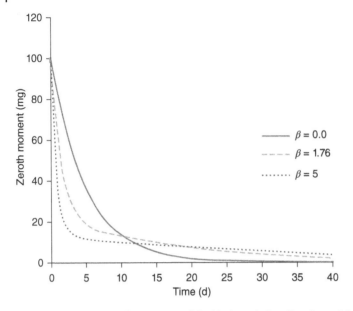

Figure 5.18 Rate-limited sorption model with degradation. Zeroth spatial moment versus time for the Table 5.2 parameter values with $\beta = 0$, 1.76, 5.0.

and the sorption distribution coefficient, k_d. Since β and k_d are linearly related, we see that both α and β have similar effects on the early time behavior of the zeroth moment versus time plot. At later times, long tails of the zeroth moment versus time plot are correlated with large values of β, since larger values of β result in more mass being "preserved" in the immobile phase (since mass in the immobile phase does not undergo degradation).

First Spatial Moment Behavior When There Is Rate-Limited Sorption and Degradation
Figure 5.19 plots the first moment of the concentration distribution versus time for different values of the first-order degradation rate constant (λ). We note that the slope of the plot indicates the velocity of the solute plume. We see from Figure 5.19b that at small times, the velocity is the pore velocity of the groundwater (v) since at small times mass transfer of the dissolved solute to the immobile (sorbed) phase is small. Here, as before, we can quantify "small" times using a Damköhler number (Da_I^s), defined as the ratio of the time to the reciprocal of the mass transfer rate constant $\left(\frac{t}{1/\alpha} \right)$. Small times may be defined as times for which $Da_I^s \ll 1$. Thus, for Table 5.2 parameter values, mass transfer of dissolved solute to the sorbed phase may be considered insignificant for times much less than 5 days.

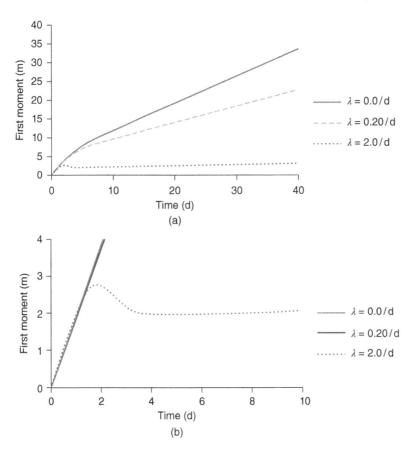

Figure 5.19 Rate-limited sorption model with degradation. (a) first spatial moment versus time for the Table 5.2 parameter values with $\lambda = 0.0, 0.20$, and $2.0\,\mathrm{d}^{-1}$, (b) detail showing behavior at small times.

We can derive an equation for the long-time $(Da_I^s \gg 1)$ behavior of the first moment by ignoring the lower order terms in Equations (5.33) and (5.34) as $t \to \infty$. Doing this, we find the following linear relationship:

$$\mu_1'^{\infty} = b_1 t + c_1 \tag{5.38}$$

where

$$b_1 = \frac{1}{2}v(1-\gamma)$$

$$c_1 = \frac{1}{q_0}v(1+\gamma)$$

and $\mu_1^{\prime\infty}$ is the mean of the concentration distribution in space calculated at long times. We see from Equation (5.38) that b_1 represents the constant velocity of the solute plume at long times and that b_1 is a function of the groundwater pore velocity (v), as well as the values of the degradation (λ) and mass transfer (α) rate constants and the affinity of the compound for the immobile (sorbed) phase (β). Figure 5.19a clearly shows how velocity is constant at long times.

Interestingly, we note at intermediate times for certain combinations of parameter values, there is a region where the slope of the concentration first moment versus time curve is negative, indicating a negative velocity (see Figure 5.19b). This is a consequence of the assumption that degradation only occurs in the dissolved phase. Thus, there is a time period where the plume, which has been moving at approximately the groundwater pore velocity, is degrading relatively rapidly, while a significant mass (relative to the dissolved mass remaining in the plume), which had been sorbed upgradient of the location of the dissolved plume's first moment, is desorbing. The net effect of these two processes is to move the mean of the concentration distribution upgradient for a period of time. We also note from Figure 5.19 that the larger the value of the first-order degradation rate constant (λ), the smaller the first moment of the concentration distribution at any given time. Again, this is a consequence of the assumption that degradation occurs in the dissolved phase only. The larger the value of the degradation rate constant, the higher the fraction of mass stored in the immobile phase and, therefore, the smaller the first moment.

Figure 5.20 shows the effect of the sorption rate constant on the behavior of the first moment over time. We again see that at short and long times, velocity

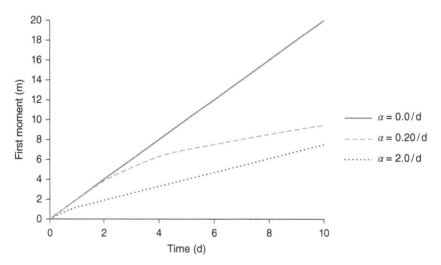

Figure 5.20 Rate-limited sorption model with degradation. First spatial moment versus time for the Table 5.2 parameter values with $\alpha = 0.0, 0.20,$ and $2.0\,\text{d}^{-1}$.

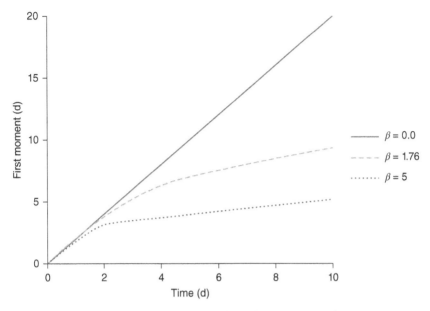

Figure 5.21 Rate-limited sorption model with degradation. First spatial moment versus time for the Table 5.2 parameter values with $\beta = 0, 1.76, 5.0$.

is constant, at values of v and b_1, respectively. We also see, as expected, that at a given time the first moment decreases with increasing values of the sorption rate constant since large α means more sorption and, therefore, more retardation of the plume.

Figure 5.21 shows that the effect of increasing β on the behavior of the first moment over time is similar to the effect of increasing α, for essentially the same reasons.

Second Spatial Moment Behavior When There Is Rate-Limited Sorption and Degradation Figure 5.22 plots the second moment of the concentration distribution versus time for different values of the first-order degradation rate constant (λ). We note that half the slope of the plot indicates the effective dispersion coefficient of the solute plume. We see from Figure 5.22b that at small times, the slope of the plot for all values of the first-order degradation rate constant is twice the dispersion coefficient (D_x) in Equation (5.32). That is, using Table 5.2 parameter values, we note that $D_x = a_x v = 0.5\,\mathrm{m}^2/d$ and we see that the slope of the second moment versus time plots at small times is $1.0\,\mathrm{m}^2/d$. As with the first moment behavior, this is because at small times mass transfer of the dissolved solute to the immobile (sorbed) phase is small. Here, as before, we can quantify "small" times using a Damköhler number (Da_I^s), and we find that for Table 5.2 parameter values, mass transfer of dissolved solute to the sorbed phase may be considered insignificant for times much less than 5 days.

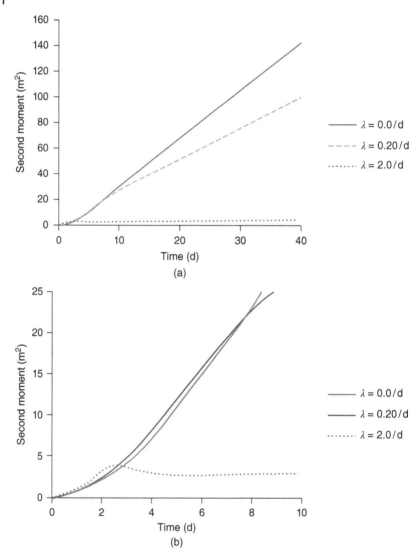

Figure 5.22 Rate-limited sorption model with degradation. (a) Second spatial moment versus time for the Table 5.2 parameter values with $\lambda = 0.0, 0.20$, and $2.0\,\mathrm{d}^{-1}$, (b) detail showing behavior at small times.

As we did for the first moment, we also can derive an equation for the long-time $(Da_I^s \gg 1)$ behavior of the second moment by ignoring the lower order terms in Equations (5.33) and (5.34) as $t \to \infty$. Doing this, we find the following linear relationship:

$$\mu_2^\infty = b_2 t + c_2 \tag{5.39}$$

where

$$b_2 = D_x(1 - \gamma) + \frac{1}{2q_0}v^2(1 - \gamma^2)$$

$$c_2 = \frac{2}{q_0}D_x(\gamma + 1) + \frac{1}{q_0^2}v^2(\gamma + 1)(2\gamma - 1)$$

and μ_2^∞ is the variance of the concentration distribution in space calculated at long times. We see from Equation (5.39) that b_2 represents twice the constant dispersion coefficient of the solute plume at long times and that b_2 is a function of the groundwater pore velocity (v) and dispersivity (a_x), as well as the values of the degradation (λ) and mass transfer (α) rate constants and the affinity of the compound for the immobile (sorbed) phase (β). Figure 5.22a clearly shows how the dispersion coefficient is constant at long times for all values of the first-order degradation rate constant. As might be expected, at long times, the greater the first-order degradation rate constant, the smaller the dispersion, since degradation decreases the extent (and spread) of the plume.

It is interesting to observe how the dispersion coefficient changes over time from its short-time value (D_x) to its long-time value ($b_2/2$). We note from Figure 5.22 that depending on parameter values, the long-time dispersion coefficient can be greater or less than the short-time value. We see from Figure 5.22b that at relatively early times (i.e., between 1 and 2 days), for all values of λ, the dispersion coefficient increases from its initial value of D_x. This may be explained by sorbed compound upgradient of the dissolved plume desorbing, resulting in an increased change in variance with time. Then, as the desorbed compound degrades, the change in variance with time begins to decrease. Interestingly, we note that this decrease in the slope of the second moment versus time curve can result in a negative slope (and, therefore, a negative dispersion coefficient) for certain combinations of parameter values (Figure 5.22b). That is, the plume is sharpening (spreading decreases) for a period of time. As with the first moment behavior, this is a consequence of the assumption that degradation only occurs in the dissolved phase. Thus, there is a time period where the solute plume, which has experienced a rapid increase in the variance with time due to desorption, now experiences an even more rapid decrease in the variance with time curve due to the relatively rapid degradation of the desorbed compound. Consistent with this explanation, we note from Figure 5.22 that the larger the value of the first-order degradation rate constant (λ), the larger the decrease in the slope of the variance versus time curve.

Figures 5.23 and 5.24 plot variance versus time for changing values of the first-order mass transfer rate constant (α) and the affinity of the compound for the immobile (sorbed) phase (β), respectively. In both figures, the dispersion coefficient goes from its short-time value (D_x) to its long-time value ($b_2/2$). From Figure 5.23, we see that for large values of the first-order mass transfer rate

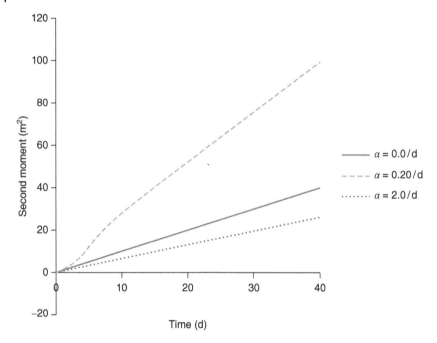

Figure 5.23 Rate-limited sorption model with degradation. Second spatial moment versus time for the Table 5.2 parameter values with $\alpha = 0.0$, 0.20, and 2.0 d^{-1}.

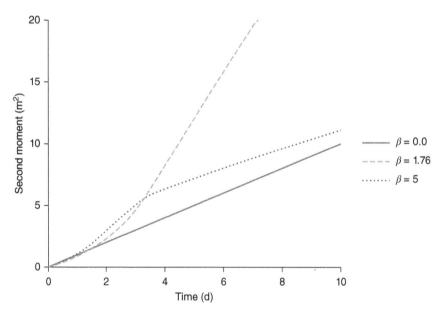

Figure 5.24 Rate-limited sorption model with degradation. Second spatial moment versus time for the Table 5.2 parameter values with $\beta = 0$, 1.76, 5.0.

constant ($\alpha \gg \lambda$), the long-time dispersion coefficient approaches the value of the long-time dispersion coefficient obtained for rate-limited sorption with no degradation (Equation (5.37)). For moderate values of the mass transfer rate constant, dispersion is increased due to the effect of rate-limited sorption on plume spreading. Similarly, Figure 5.24 shows that moderate values of β result in enhanced dispersion compared to small and large values of β. Note from Equation (5.39) that as β gets very large, dispersion approaches 0 (since mass is all associated with the immobile phase).

Problems

5.1 Using formulas presented in this chapter, calculate the zeroth absolute moment, the first normalized absolute moment, and the second moment about the mean of a breakthrough curve at $x = 2$ m that would result from the system described by Equation (5.6) with boundary and initial conditions (5.7) and the Table 5.1 parameters.

5.2 Which AnaModelTool model is appropriate to use to simulate Equation (5.6) with boundary and initial conditions (5.7)?

5.3 Using the model identified in Problem 5.2, simulate a breakthrough curve at $x = 2$ m for the Table 5.1 parameter values. Download the breakthrough data into a spreadsheet and numerically evaluate the zeroth, first, and second moments of the breakthrough curve that was generated. Compare the moments that were obtained numerically with the moments that were calculated in Problem 5.1; they should be approximately equal. HINT: To evaluate the moments of the breakthrough curve, it will be necessary to numerically evaluate the integrals in Equation (5.1) for $j = 0, 1$, and 2 using a quadrature method such as the trapezoidal or Simpson's rules.

5.4 Use Models 104 and 104M with the Table 5.2 parameter values to simulate the spatial concentration distributions and moments of a slowing sorbing, nondegrading ($\lambda = 0$) contaminant in space at very short and very long times (compared to $1/\alpha$). Then use Equations (5.36) and (5.37) to calculate effective dispersion coefficients at short and long times and use those values in Model 104 and 104M to simulate the concentration distributions and moments of an instantaneously sorbing, nondegrading ($\lambda = 0$) contaminant in space at short and long times. Compare the short- and long-time moments and distributions (rate-limited vs equilibrium) and discuss in terms of symmetry, mass, center of mass, and spreading.

References

Aris, R. (1958), On the dispersion of linear kinematic waves, *Proceedings of the Royal Society of London, Series A, 245*, 268–277.

Domenico, P.A. and F.W. Schwartz, *Physical and Chemical Hydrogeology*, 2nd Edition, Wiley, New York, 1998.

Goltz, M.N. and P.V. Roberts, Using the method of moments to analyze three-dimensional diffusion-limited solute transport from temporal and spatial perspectives, *Water Resources Research, 23*(8):1575–1585, 1987.

Govindaraju, R.S. and B.S. Das, *Moment Analysis for Subsurface Hydrologic Applications, Springer*, The Netherlands, 2007.

6

Application of Analytical Models to Gain Insight into Transport Behavior

In this chapter, we apply our models to simulate situations that might be encountered in the "real world." Of course, due to the simplifications that we need to make to analytically solve the models, we are very far from simulating real-world conditions. That said, however, the model results may be used to provide insights into transport behavior that will be helpful to decision-makers who are dealing with groundwater contamination problems.

6.1 Contaminant Remediation

Let us consider the scenario depicted in Figure 1.5. An extraction well of radius r_w starts pumping at a flow Q_w at time zero. The well is at the center of a cylindrical contaminated zone of radius r_b and height b. Contaminant transport in the aquifer is affected by radial advection, dispersion, and rate-limited sorption, modeled as a first-order process. Goltz and Oxley (1991) modeled this scenario and developed Laplace domain solutions to the model equations for both concentration at the extraction well and total mass remaining in the aquifer as functions of time. For the interested reader, the model equations are presented and the Laplace domain solutions derived in Appendix M. These solutions are used in AnaModelTool Model 403.

Baseline parameter values for a remediation scenario of a nondegrading contaminant (first-order degradation rate constant, λ, equals zero) are listed in Table 6.1. Let us examine how changes in the parameter values affect aquifer cleanup. We will define cleanup as the achievement of concentrations at the extraction well that are less than 0.005 of the initial contaminant concentration. This is equivalent to cleaning up an aquifer that has contaminant concentrations of 1 mg/L to a level of 5 μg/L, which is the regulatory maximum contaminant level mandated in Safe Drinking Water Act regulations for a number of carcinogenic compounds that are frequently found in groundwater.

The sensitivity analysis in Table 6.2 shows a number of expected results. We see, as expected, that the cleanup time is approximately inversely proportional

Analytical Modeling of Solute Transport in Groundwater: Using Models to Understand the Effect of Natural Processes on Contaminant Fate and Transport, First Edition. Mark Goltz and Junqi Huang.
© 2017 John Wiley & Sons, Inc. Published 2017 by John Wiley & Sons, Inc.
Companion Website: www.wiley.com/go/Goltz/solute_transport_in_groundwater

Table 6.1 Baseline parameter values for remediation sensitivity analysis.

Parameter	Value
Extraction well pumping rate (Q_w)	20 m³/d
Radius of contaminated zone (r_b)	50 m
Extraction well radius (r_w)	0.1 m
Aquifer thickness (b)	5 m
Porosity (θ)	0.25
Retardation factor (R)	2.0
Dispersivity (a_r)	0.1 m
First-order mass transfer rate constant (α)	0.5 d⁻¹
Initial concentration in contaminated zone (C_0)	1 mg/L

to the extraction well pumping rate; directly proportional to the aquifer thickness, porosity, and retardation factor; and proportional to the square of the contaminated zone radius. We also note that as would be guessed, the radius of the extraction well does not impact cleanup time, since it is the extraction well pumping rate, not its radius, that impacts cleanup time. Also as would be anticipated, increased dispersivities result in increased cleanup times, since more spreading means longer contaminant concentration "tails" that cause the concentrations to remain above regulatory limits for a longer time.

The relationship between cleanup time and the sorption rate constant is quite interesting, and perhaps not anticipated. We see that at very low and very high values of the rate constant, cleanup times are relatively short, while for intermediate rate constant values, cleanup times are increased dramatically. Let us first think about what happens at very high values of the rate constant. In essence, when the sorption rate constant is large (in comparison to the advection rate constant), sorption may be assumed to be at equilibrium. For the example scenario described by the Table 6.1 parameters, the advection rate constant (r_a) can be calculated as (Goltz and Oxley, 1991)

$$r_a = \frac{v_w(r_b)}{r_b R} = \frac{Q_w/(2\pi r_b b\theta)}{r_b R} = 0.0005 \, \text{d}^{-1}$$

where $v_w(r_b)$ is the pore velocity of the groundwater at the boundary of the contaminated zone. We see that when the sorption rate constant is much larger than r_a, sorption approaches equilibrium and the cleanup time is the same as the cleanup time that is obtained when equilibrium sorption is assumed (i.e., for this case, 1185 d when $\alpha \to \infty$).

At very low values of the sorption rate constant (low in comparison to r_a), sorption is so slow that it can be ignored. Thus, we see that cleanup times for

Table 6.2 Effect of parameter value on cleanup time.[a]

Parameter	Value	Cleanup time (d)
Extraction well pumping rate (Q_w)	10 m^3/d	2400
	20 m^3/d	1215
	40 m^3/d	625
Radius of contaminated zone (r_b)	25 m	340
	50 m	1215
	100 m	4530
Extraction well radius (r_w)	0.01 m	1215
	0.1 m	1215
	1.0 m	1215
Aquifer thickness (b)	1 m	265
	5 m	1215
	10 m	2400
Porosity (θ)	0.125	620
	0.25	1215
	0.50	2400
Retardation factor (R)	1.0	590
	2.0	1215
	4.0	2415
Dispersivity (a_r)	0.1 m	1215
	1.0 m	1730
	10.0 m	3510
First-order mass transfer rate constant (α)	∞	1185
	>5.0 d^{-1}	1185
	0.5 d^{-1}	1215
	0.05 d^{-1}	1443
	0.0005 d^{-1}	8970
	0.000005 d^{-1}	32,460
	0.000001 d^{-1}	640
	<1e−7 d^{-1}	590

a) Cleanup time defined as the time needed to reduce concentrations at the extraction well to 0.005 times the initial concentration.

$\alpha < 1e - 7 \ d^{-1}$ are the same as for when there is no sorption and the retardation factor is unity.

As we saw in Chapter 3, when the sorption rate constant is comparable to r_a, contaminant concentration tailing is very significant and cleanup times are orders of magnitude greater than when the sorption rate is either very slow or very fast.

6.2   Borden Field Experiment

Beginning in 1982, a large-scale field experiment was conducted at Canadian Forces Base (CFB) Borden to study the fate and transport in groundwater of dissolved halogenated compounds. Two conservative (nonsorbing, nondegrading) tracers, chloride and bromide, along with five halogenated compounds (carbon tetrachloride, tetrachloroethene, 1,2-dichlorobenzene, bromoform, and hexachloroethane) were injected into groundwater that was moving at approximately 9.1 cm/d. The groundwater pore velocity that was measured agreed closely with that calculated using the hydraulic gradient, hydraulic conductivity, and porosity of the aquifer as input parameters for Darcy's Law.

An extensive three-dimensional monitoring network was installed to track the movement of the seven individual contaminant plumes over space and time. Two types of sampling were conducted: (1) temporal or "breakthrough" sampling, where concentrations as a function of time were measured at specific locations, and (2) synoptic or "snap shot" sampling, where concentrations were obtained throughout the entire monitoring network at specified times. Data from the temporal sampling were used to plot concentration versus time

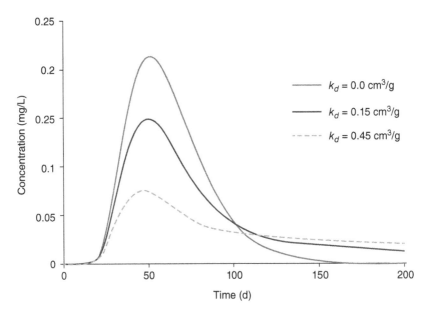

Figure 6.1 Model 104 qualitative simulation of CFB Borden experiment breakthrough curves obtained at a monitoring well 500 cm from the injection zone for a conservative tracer ($k_d = 0.0$ cm³/g) and two sorbing but nondegrading chemicals ($k_d = 0.15$ cm³/g and $k_d = 0.45$ cm³/g). Table 6.3 parameter values.

breakthrough curves, while data from the synoptic sampling sessions were used to calculate the zeroth, first, and second moments of the concentration distributions in space.

The temporal sampling showed that for nondegrading compounds (in addition to the two tracers, carbon tetrachloride and tetrachloroethene were found to not degrade in the aerobic aquifer) the greater the sorption, the greater the breakthrough curve tailing. Thus, when comparing the breakthrough curves of chloride/bromide ($k_d = 0.0$ cm^3/g), carbon tetrachloride, which had a measured k_d of 0.15 cm^3/g (Curtis *et al.*, 1986), and tetrachloroethene, which had a measured k_d of 0.45 cm^3/g (Curtis *et al.*, 1986), it was seen that the breakthrough curve tailing was most significant for tetrachloroethene, less significant for carbon tetrachloride, and insignificant for chloride/bromide.

We saw in Chapter 3 that rate-limited sorption may result in breakthrough curve tailing. Let us now examine whether what was observed in the field at CFB Borden was predicted by our models; that is, for the same value of the first-order mass transfer rate constant (α), increasing values of the sorption distribution coefficient (k_d) result in increased tailing.

Table 6.3 shows the parameters that were used as input for Model 104 to qualitatively simulate the concentration versus time breakthrough results observed in the field at CFB Borden at a monitoring well 500 cm downgradient of the injection zone for the conservative tracer and the two sorbing but nondegrading compounds (carbon tetrachloride and tetrachloroethene). The simulation is qualitative only, since we are using a one-dimensional model to simulate a three-dimensional system. Nevertheless, we see from Figure 6.1 that indeed, as the sorption distribution coefficient increases, the degree of breakthrough curve tailing increases, as was observed in the field.

We now assess the synoptic sampling results that were observed at CFB Borden in light of our models. Let us consider the behavior of the zeroth, first, and

Table 6.3 Parameters (based on CFB Borden Study) used in model 104 to examine the effect of the sorption distribution coefficient on breakthrough curve tailing.

Parameter	Value
Porosity (θ)	0.33
Bulk density (ρ_b)	1.81 g/cm^3
Dispersion coefficient (D_x)	330 cm^2/d
Pore velocity (v)	9.1 cm/d
Degradation rate constant (λ)	0.0 d^{-1}
First-order mass transfer rate constant (α)	0.01 d^{-1}
Sorption distribution coefficient (k_d)	0.0, 0.15, 0.45 cm^3/g
Injected mass (M)	100 g

second moments of the concentration distributions in space if sorption were assumed to be at equilibrium (i.e., sorption is fast compared to other processes affecting chemical fate and transport). From the discussion in Section 5.2.3.1, we expect that for nondegrading chemicals, the zeroth moment would be a constant (see Equation (5.29)), while the first and second moments would be linear functions of time (Equations (5.30) and (5.31), respectively). In fact, what was observed from the synoptic sampling at CFB Borden was that for the sorbing nondegrading chemicals (carbon tetrachloride and tetrachloroethene), the zeroth moment declined over time, ultimately approaching a constant value, while plots of the first and second moments were not linear functions of time. It was seen that a plot of the first moment versus time had a decreasing slope (indicating deceleration of the plume with time), with the slope ultimately approaching a constant value, while a plot of the second moment versus time had an increasing slope (indicating increased spreading over time), with the slope ultimately approaching a constant value (Goltz, 1986; Goltz and Roberts, 1988). Going back to the discussion in Section 5.2.3.2.1 and Figures 5.13–5.15, we see this is precisely the spatial moment behavior predicted by a model that simulates rate-limited sorption of a nondegrading chemical.

Also at CFB Borden, the synoptic sampling of the conservative tracers, chloride and bromide, showed the expected spatial moment behavior. That is, the zeroth moment was constant over time and the first and second moments were linear functions of time (Freyberg, 1986).

Problems

6.1 The sensitivity analysis results reported in Table 6.2 are based on a nondegrading contaminant. Using the baseline parameter values in Table 6.1, Run Model 403 with small, moderate, and high values of the Damköhler number (0.01, 1.0, and 100, respectively) to determine cleanup times. To calculate the first-order degradation rate constant from the Damköhler number, use the reciprocal of the advection rate constant (r_a) as the advection timescale. Discuss the relationship between Damköhler number and cleanup time.

6.2 Use Model 104M and the parameter values in Table 6.3 (with $k_d = 0.45$ cm^3/g) to plot zeroth, first, and second spatial moments as a function of time. From the plots, determine the long-time values of mass, pore velocity, and dispersion. Compare these values with values calculated from the equations in Section 5.2.3.2.1.

References

Curtis, G.P., P.V. Roberts, and M. Reinhard, A natural gradient experiment on solute transport in a sand aquifer: 4. Sorption of organic solutes and its influence on mobility, *Water Resources Research, 22*(13): 2059–2067, 1986.

Freyberg, D.L., A natural gradient experiment on solute transport in a sand aquifer: 2. Spatial moments and the advection and dispersion of nonreactive tracers. *Water Resources Research, 22*(13): 2031–2046, 1986.

Goltz, M.N., Three-dimensional analytical modeling of diffusion-limited solute transport, Ph.D. Thesis, Stanford University, Stanford, CA, 172 pp., 1986.

Goltz, M.N. and M.E. Oxley, Analytical modeling of aquifer decontamination by pumping when transport is affected by rate-limited sorption, *Water Resources Research, 27*(4): 547–556, 1991.

Goltz, M.N. and P.V. Roberts, Simulations of physical nonequilibrium solute transport models: application to a large-scale field experiment, *Journal of Contaminant Hydrology, 3*: 37–63, 1988.

A

Solution to One-Dimensional ADR Equation with First-Order Degradation Kinetics Using Laplace Transforms

In this appendix, we solve the one-dimensional ADR equation as formulated in Equation (2.21a) using first-order degradation kinetics. We assume that the system is initially contaminated at a concentration C_i (initial condition (2.24)), a first-type boundary condition at $x = 0$ (boundary condition (2.29)) and a second-type boundary condition at $x = \infty$ (Equation (2.30)). Rewriting the ADR with its initial and boundary conditions (IC/BCs) gives

$$\frac{\partial C}{\partial t} = -\frac{v}{R}\frac{\partial C}{\partial x} + \frac{D_x}{R}\frac{\partial^2 C}{\partial x^2} - \frac{1}{R}\lambda C \tag{A.1}$$

$$C(x, t = 0) = C_i \tag{A.2a}$$

$$C(x = 0, t) = C_0 \tag{A.2b}$$

$$\frac{\partial C}{\partial x}(x = \infty, t) = 0 \tag{A.2c}$$

Our approach will be to use Laplace transforms to convert the PDE with its IC/BCs into an ordinary differential equation (ODE) with BCs in "Laplace time" that can be solved using standard methods. We then will invert the Laplace time solution back into "real time" to obtain the solution for concentration as a function of x and t. Appendix E is a table that includes useful Laplace transforms.

To begin, we must define the Laplace transform of any function of time $f(t)$ as follows:

$$\Im\{f(t)\} = \bar{f}(s) = \int_0^\infty e^{-st} f(t)dt$$

where s is the Laplace transform variable. The Laplace transform has a very useful property, in that in Laplace time, the time derivative is transformed to a function of s. That is

$$\Im\{f'(t)\} = s\bar{f}(s) - f(0)$$

This property allows us to convert a PDE in space and time, to an ODE in space only, since the time derivative is transformed into a function of the

Analytical Modeling of Solute Transport in Groundwater: Using Models to Understand the Effect of Natural Processes on Contaminant Fate and Transport, First Edition. Mark Goltz and Junqi Huang.
© 2017 John Wiley & Sons, Inc. Published 2017 by John Wiley & Sons, Inc.
Companion Website: www.wiley.com/go/Goltz/solute_transport_in_groundwater

Laplace transform variable. With the following definitions,

$$\Im\{C(x,t)\} = \overline{C}(x,s) = \int_0^\infty e^{-st} C(x,t)dt$$

$v' = \frac{v}{R}, D' = \frac{D_x}{R}, \lambda' = \frac{\lambda}{R}$, we apply the Laplace transformation to Equation (A.1) and IC/BCs (A.2) to obtain the following ODE with BCs:

$$s\overline{C}(x,s) - C(x,t=0) = -v'\frac{d\overline{C}}{dx} + D'\frac{d^2\overline{C}}{dx^2} - \lambda'\overline{C} \tag{A.3}$$

$$\overline{C}(x=0,s) = \frac{C_0}{s} \tag{A.4a}$$

$$\frac{d\overline{C}}{dx}(x=\infty,p) = 0 \tag{A.4b}$$

Note how the Laplace transformation eliminated the time derivative on the left-hand side of Equation (A.1), thereby transforming the PDE into an ODE (Equation (A.3)). Making use of the IC, which allows us to substitute C_i for the second term on the left-hand side of Equation (A.3), we rearrange the terms in Equation (A.3) to obtain the following linear, nonhomogeneous, second-order ODE with constant coefficients:

$$\frac{d^2\overline{C}}{dx^2} - \frac{v'}{D'}\frac{d\overline{C}}{dx} - \frac{s+\lambda'}{D'}\overline{C} = \frac{-C_i}{D'} \tag{A.5}$$

The general solution to this equation may be found in any text on ODEs:

$$\overline{C}(x,s) = C_1 e^{\left(\frac{v'}{2D'}+\frac{1}{2}\sqrt{\left(\frac{v'}{D'}\right)^2+\frac{4(s+\lambda')}{D'}}\right)x} + C_2 e^{\left(\frac{v'}{2D'}-\frac{1}{2}\sqrt{\left(\frac{v'}{D'}\right)^2+\frac{4(s+\lambda')}{D'}}\right)x} + \frac{C_i}{s+\lambda'} \tag{A.6}$$

To obtain the particular solution, we must apply BCs (A.4). Since the exponent of the first-term on the right-hand side of Equation (A.6) is positive, we know that the derivative with respect to x of this term will also have a positive exponent. Thus, the coefficient C1 must be zero in order to satisfy BC (A-4b), as otherwise the term would be infinite at $x = \infty$. To satisfy BC (A.4a), we see $C_2 = \frac{C_0}{s} - \frac{C_i}{s+\lambda'}$. Thus, the particular solution to ODE (A.5) with BCs (A.4) is

$$\overline{C}(x,s) = \frac{C_0}{s} e^{\left(\frac{v'}{2D'}-\frac{1}{2}\sqrt{\left(\frac{v'}{D'}\right)^2+\frac{4(s+\lambda')}{D'}}\right)x} - \frac{C_i}{s+\lambda'} e^{\left(\frac{v'}{2D'}-\frac{1}{2}\sqrt{\left(\frac{v'}{D'}\right)^2+\frac{4(s+\lambda')}{D'}}\right)x} + \frac{C_i}{s+\lambda'} \tag{A.7}$$

Solution (A.7) is in Laplace time. In order to obtain the solution in real time, we must invert it. The Laplace inversion formula is

$$C(x,t) = \frac{1}{2\pi i}\int_{c-i\infty}^{c+i\infty} \overline{C}(x,s)e^{st} dt \tag{A.8}$$

where c is a number on the real axis, which is chosen such that any singularities are to the left of the line passing through $(c, 0i)$ and parallel to the imaginary axis. Note that inverse Laplace transforms are linear, so we can invert each of the three terms in the Laplace time solution individually to obtain the real-time solution.

The simplest way to invert Laplace time solution (A.7) is to use Laplace inversion tables. Defining $a = \frac{v'^2}{4D'} + \lambda'$, we may rearrange the first term on the right-hand side of Equation (A.7) to put it in a form that is similar to a tabulated form:

$$C_0 e^{\frac{v'x}{2D'}} \frac{e^{\frac{-x}{\sqrt{D'}}\sqrt{s+a}}}{(s+a)-a} \tag{A.9}$$

Noting that another property of Laplace transforms is that the Laplace inversion of $\overline{C}(x, s+a)$ is $e^{-at}C(x, t)$, we use Appendix E to find the inverse of (A.9) is

$$C_0 e^{\frac{v'x}{2D'}} e^{-at} \frac{1}{2} \left[e^{at-\frac{x}{\sqrt{D'}}\sqrt{a}} \operatorname{erfc}\left(\frac{x}{2\sqrt{D't}} - \sqrt{at} \right) \right.$$
$$\left. + e^{at+\frac{x}{\sqrt{D'}}\sqrt{a}} \operatorname{erfc}\left(\frac{x}{2\sqrt{D't}} + \sqrt{at} \right) \right] \tag{A.10}$$

The second term on the right-hand side of Equation (A.7) can be rewritten as follows:

$$- C_i e^{\frac{v'x}{2D'}} \frac{e^{\frac{-x}{\sqrt{D'}}\sqrt{s+a}}}{(s+a)-\frac{(v')^2}{4D'}} \tag{A.11}$$

and by again using the Laplace transform table in Appendix E, we find that the inverse of (A.11) is

$$-C_i e^{\frac{v'x}{2D'}} e^{-at} \frac{1}{2} \left[e^{\frac{(v')^2 t}{4D'}-\frac{v'x}{2D'}} \operatorname{erfc}\left(\frac{x-v't}{2\sqrt{D't}} \right) + e^{\frac{(v')^2 t}{4D'}+\frac{v'x}{2D'}} \operatorname{erfc}\left(\frac{x+v't}{2\sqrt{D't}} \right) \right]$$

or simplified as

$$- \frac{C_i}{2} e^{-\lambda't} \left[\operatorname{erfc}\left(\frac{x-v't}{2\sqrt{D't}} \right) + e^{\frac{vx}{D'}} \operatorname{erfc}\left(\frac{x+v't}{2\sqrt{D't}} \right) \right] \tag{A.12}$$

The Laplace transform table shows the inverse of the third term on the right-hand side of Equation (A.7) is

$$e^{-\lambda't} C_i \tag{A.13}$$

Substituting for λ', D', v', and a, based on their definitions, and combining terms (A.10), (A.12), and (A.13), we can write the fully inverted expression for

Equation (A.7) as

$$C(x,t) = \frac{1}{2}C_0 \left\{ e^{\frac{(v-u)x}{2D_x}} \operatorname{erfc}\left[\frac{Rx - ut}{2\sqrt{D_x Rt}}\right] + e^{\frac{(v+u)x}{2D_x}} \operatorname{erfc}\left[\frac{Rx + ut}{2\sqrt{D_x Rt}}\right] \right\}$$
$$- \frac{1}{2}C_i e^{-\frac{\lambda t}{R}} \left\{ \operatorname{erfc}\left[\frac{Rx - vt}{2\sqrt{D_x Rt}}\right] + e^{\frac{vx}{D_x}} \operatorname{erfc}\left[\frac{Rx + vt}{2\sqrt{D_x Rt}}\right] \right\} + C_i e^{-\frac{\lambda t}{R}}$$

$$(A.14)$$

where $u = v\sqrt{1 + \frac{4\lambda D_x}{v^2}}$ and erfc is the complementary error function. This solution may also be found in van Genuchten and Alves (1982).

Reference

van Genuchten, M.Th. and W.J. Alves, *Analytical Solutions to the One-Dimensional Convective-Dispersive Solute Transport Equation*, U.S. Department of Agriculture, Technical Bulletin No. 1661, 151 pp., 1982.

B

Solution to One-Dimensional ADR Equation with Zeroth-Order Degradation Kinetics Using Laplace Transforms

In this appendix, we solve the one-dimensional ADR equation as formulated in Equation (2.21a) using zeroth-order degradation kinetics. We assume the system is initially contaminated at a concentration C_i (initial condition (2.24)), a first-type boundary condition at $x = 0$ (boundary condition (2.29)) and a second-type boundary condition at $x = \infty$ (BC (2.30)). Rewriting the ADR with its initial and boundary conditions (IC/BCs) gives

$$\frac{\partial C}{\partial t} = -\frac{v}{R}\frac{\partial C}{\partial x} + \frac{D_x}{R}\frac{\partial^2 C}{\partial x^2} - \frac{1}{R}k_0 \tag{B.1}$$

$$C(x, t = 0) = C_i \tag{B.2a}$$

$$C(x = 0, t) = C_0 \tag{B.2b}$$

$$\frac{\partial C}{\partial x}(x = \infty, t) = 0 \tag{B.2c}$$

As in Appendix A, our approach will be to use Laplace transforms to convert the PDE with its IC/BCs into an ordinary differential equation (ODE) with BCs in "Laplace time" that can be solved using standard methods. We then will invert the Laplace time solution back into "real time" to obtain the solution for concentration as a function of x and t.

Following Appendix A, we define

$$v' = \frac{v}{R}, D' = \frac{D_x}{R}, k_0' = \frac{k_0}{R}$$

and apply the Laplace transformation to Equation (B.1) and IC/BCs (B.2) to obtain the following ODE with BCs:

$$s\overline{C}(x, s) - C(x, t = 0) = -v'\frac{d\overline{C}}{dx} + D'\frac{d^2\overline{C}}{dx^2} - \frac{k_0'}{s} \tag{B.3}$$

Analytical Modeling of Solute Transport in Groundwater: Using Models to Understand the Effect of Natural Processes on Contaminant Fate and Transport, First Edition. Mark Goltz and Junqi Huang.
© 2017 John Wiley & Sons, Inc. Published 2017 by John Wiley & Sons, Inc.
Companion Website: www.wiley.com/go/Goltz/solute_transport_in_groundwater

$$\overline{C}(x=0,s) = \frac{C_0}{s} \tag{B.4a}$$

$$\frac{d\overline{C}}{dx}(x=\infty,s) = 0 \tag{B.4b}$$

Rearranging terms in Equation (B.3) results in the following linear, nonhomogeneous, second-order ODE with constant coefficients:

$$\frac{d^2\overline{C}}{dx^2} - \frac{v'}{D'}\frac{d\overline{C}}{dx} - \frac{s}{D'}\overline{C} = \frac{k'_0}{D's} - \frac{C_i}{D'} \tag{B.5}$$

The general solution to this equation may be found in an ODE text as

$$\overline{C}(x,s) = C_1 e^{\left(\frac{v'}{2D'}+\frac{1}{2}\sqrt{\left(\frac{v'}{D'}\right)^2+\frac{4s}{D'}}\right)x} + C_2 e^{\left(\frac{v'}{2D'}-\frac{1}{2}\sqrt{\left(\frac{v'}{D'}\right)^2+\frac{4s}{D'}}\right)x} - \frac{k'_0}{s^2} + \frac{C_i}{s} \tag{B.6}$$

To obtain the particular solution, we must apply BCs (B.4). Since the exponent of the first-term on the right-hand side of Equation (B.6) is positive, we know that the derivative with respect to x of this term will also have a positive exponent. Thus, the coefficient C1 must be zero in order to satisfy BC (A-4b), as otherwise the term would be infinite at $x = \infty$. To satisfy BC (B.4a), we see $C_2 = \frac{C_0}{s} + \frac{k'_0}{s^2} - \frac{C_i}{s}$. Thus, the particular solution to ODE (B.5) with BCs (B.4) is

$$\overline{C}(x,s) = \frac{C_0 - C_i}{s} e^{\left(\frac{v'}{2D'}-\frac{1}{2}\sqrt{\left(\frac{v'}{D'}\right)^2+\frac{4s}{D'}}\right)x} + \frac{k'_0}{s^2} e^{\left(\frac{v'}{2D'}-\frac{1}{2}\sqrt{\left(\frac{v'}{D'}\right)^2+\frac{4s}{D'}}\right)x} - \frac{k'_0}{s^2} + \frac{C_i}{s} \tag{B.7}$$

We now want to invert Laplace time solution (B.7) to obtain the real-time solution. As in Appendix A, we note that inverse Laplace transforms are linear, so we can invert each term in the Laplace time solution individually to obtain the real-time solution.

Noting that the first term on the right-hand side of Equation (B.7) is identical in form to the second term on the right-hand side of Equation (A.7), we apply the methods of Appendix A to find the inverse Laplace transform of the first term on the right-hand side of Equation (B.7):

$$\frac{1}{2}(C_0 - C_i)\left\{ \text{erfc}\left[\frac{x-v't}{2\sqrt{D't}}\right] + e^{\frac{v'x}{D'}}\text{erfc}\left[\frac{x+v't}{2\sqrt{D't}}\right]\right\} \tag{B.8}$$

Using tables of inverse Laplace transforms, we immediately find that the inverse of the third and fourth terms on the right-hand side of Equation (B.7) are $-k'_0 t$ and C_i, respectively. Inverting the second term on the right-hand side of Equation (B.7) requires us to put the term in a form similar to a tabulated form. Defining a as $a = \frac{v'^2}{4D'}$, we may rearrange the second term to write

$$k'_0 e^{\frac{v'x}{2D'}} \frac{e^{-\frac{x}{\sqrt{D'}}\sqrt{s+a}}}{(s+a-a)^2} \tag{B.9}$$

Using the fact that the Laplace inversion of $\overline{C}(x, s + a)$ is $e^{-at}C(x, t)$, we can use the Laplace transform table in van Genuchten and Alves (1982) to find the inverted form of the second term on the right-hand side of Equation (B.7):

$$
k_0' e^{\frac{v'x}{2D'}} e^{-at} \left[e^{at - \frac{v'x}{2D'}} \operatorname{erfc}\left(\frac{x}{2\sqrt{D't}} - \sqrt{at} \right)\left(\frac{t}{2} - \frac{x}{2v'} \right) \right.
$$

$$
\left. + e^{at + \frac{v'x}{2D'}} \operatorname{erfc}\left(\frac{x}{2\sqrt{D't}} + \sqrt{at} \right)\left(\frac{t}{2} + \frac{x}{2v'} \right) \right] \tag{B.10}
$$

Substituting for k_0', D', v', and a, based on their definitions, we combine the four inverted terms to obtain

$$
C(x, t) = \frac{1}{2}(C_0 - C_i) \left\{ \operatorname{erfc}\left[\frac{Rx - vt}{2\sqrt{D_xRt}} \right] + e^{\frac{vx}{D_x}} \operatorname{erfc}\left[\frac{Rx + vt}{2\sqrt{D_xRt}} \right] \right\}
$$

$$
+ \frac{k_0}{R}\left(\frac{vt - Rx}{2v} \right) \operatorname{erfc}\left(\frac{Rx - vt}{2\sqrt{D_xRt}} \right)
$$

$$
+ \frac{k_0}{R} e^{\frac{vx}{D_x}} \left(\frac{vt + Rx}{2v} \right) \operatorname{erfc}\left(\frac{Rx + vt}{2\sqrt{D_xRt}} \right) - \frac{k_0}{R}t + C_i \tag{B.11}
$$

or

$$
C(x, t) = \left\{ \frac{1}{2}\operatorname{erfc}\left[\frac{Rx - vt}{2\sqrt{D_xRt}} \right] \right\}\left\{ (C_0 - C_i) + \frac{k_0}{R}\left(\frac{vt - Rx}{v} \right) \right\}
$$

$$
+ \left\{ \frac{1}{2}e^{\frac{vx}{D_x}}\operatorname{erfc}\left[\frac{Rx + vt}{2\sqrt{D_xRt}} \right] \right\}\left\{ (C_0 - C_i) + \frac{k_0}{R}\left(\frac{vt + Rx}{v} \right) \right\}
$$

$$
- \frac{k_0}{R}t + C_i
$$

Recall the discussion in Section 2.2.4 on zeroth-order kinetics, and note that this analytical solution allows for negative concentrations, depending on the relative values of the parameters (i.e., when the next-to-last term on the right-hand side of Equation (B.11) is large, relative to the other terms).

Reference

van Genuchten, M.Th., and W.J. Alves, *Analytical Solutions to the One-Dimensional Convective-Dispersive Solute Transport Equation*, U.S. Department of Agriculture, Technical Bulletin No. 1661, 151 pp., 1982.

C

Solutions to the One-Dimensional ADR in Literature

$$\frac{\partial C}{\partial t} = -v\frac{\partial C}{\partial x} + D_x\frac{\partial^2 C}{\partial x^2} \pm \frac{\partial C}{\partial t}_{rxn} - \frac{\rho_b}{\theta}\frac{\partial S}{\partial t}$$

Analytical Modeling of Solute Transport in Groundwater: Using Models to Understand the Effect of Natural Processes on Contaminant Fate and Transport, First Edition. Mark Goltz and Junqi Huang.
© 2017 John Wiley & Sons, Inc. Published 2017 by John Wiley & Sons, Inc.
Companion Website: www.wiley.com/go/Goltz/solute_transport_in_groundwater

Initial condition $(t=0)$	Boundary condition 1	Boundary condition 2	$\dfrac{\partial C}{\partial t}_{rxn}$	$\dfrac{\partial S}{\partial t}$	Reference
$C = C_i$	$\begin{aligned}C(0,t) &= C_0 \quad 0 < t \le t_s \\ C(0,t) &= 0 \quad t > t_s\end{aligned}$	$\left.\dfrac{\partial C}{\partial x}\right|_{x=\infty} = 0$	k_0	$k_d \dfrac{\partial C}{\partial t}$	van Genuchten and Alves (1982, p. 31)
$C = C_i$	$\begin{aligned}C(0,t) &= C_0 \quad 0 < t \le t_s \\ C(0,t) &= 0 \quad t > t_s\end{aligned}$	$\left.\dfrac{\partial C}{\partial x}\right|_{x=\infty} = 0$	k_0	$\alpha(k_d C - S)$	Toride et al. (1993)
$C = C_i$	$\begin{aligned}\left(-D_x \dfrac{\partial C}{\partial x} + vC\right)_{x=0} &= vC_0 \quad 0 < t \le t_s \\ \left(-D_x \dfrac{\partial C}{\partial x} + vC\right)_{x=0} &= 0 \quad t > t_s\end{aligned}$	$\left.\dfrac{\partial C}{\partial x}\right|_{x=\infty} = 0$	k_0	$k_d \dfrac{\partial C}{\partial t}$	van Genuchten and Alves (1982) pp. 32
$C = C_i$	$\begin{aligned}\left(-D_x \dfrac{\partial C}{\partial x} + vC\right)_{x=0} &= vC_0 \quad 0 < t \le t_s \\ \left(-D_x \dfrac{\partial C}{\partial x} + vC\right)_{x=0} &= 0 \quad t > t_s\end{aligned}$	$\left.\dfrac{\partial C}{\partial x}\right|_{x=\infty} = 0$	k_0	$\alpha(k_d C - S)$	Toride et al. (1993)
$C = C_i$	$\begin{aligned}C(0,t) &= C_0 \quad 0 < t \le t_s \\ C(0,t) &= 0 \quad t > t_s\end{aligned}$	$\left.\dfrac{\partial C}{\partial x}\right|_{x=L} = 0$	λC	$k_d \dfrac{\partial C}{\partial t}$	van Genuchten and Alves (1982, p. 63)
$C = C_i$	$\begin{aligned}C(0,t) &= C_0 \quad 0 < t \le t_s \\ C(0,t) &= 0 \quad t > t_s\end{aligned}$	$\left.\dfrac{dC}{dx}\right|_{x=\infty} = C(\infty,t) = 0$	λC	$k_d \dfrac{\partial C}{\partial t}$	van Genuchten and Alves (1982, p. 60)
$C = C_i$	$\begin{aligned}C(0,t) &= C_0 \quad 0 < t \le t_s \\ C(0,t) &= 0 \quad t > t_s\end{aligned}$	$\left.\dfrac{dC}{dx}\right|_{x=\infty} = C(\infty,t) = 0$	λC	$\alpha(k_d C - S)$	Toride et al. (1993)

Initial condition ($t=0$)	Boundary condition 1	Boundary condition 2	$\dfrac{\partial C}{\partial t}\,_{rxn}$	$\dfrac{\partial S}{\partial t}$	Reference
$C = M\delta(x)$	$\left.\dfrac{\partial C}{\partial x}\right|_{x=\infty} = 0$	$\left.\dfrac{\partial C}{\partial x}\right|_{x=-\infty} = 0$	0	$\alpha(k_d C - S)$	Goltz and Roberts (1986)
$C = C_i$	$\left(-D_x\dfrac{\partial C}{\partial x} + vC\right)_{x=0} = vC_0 \quad 0 < t \le t_s$ $\left(-D_x\dfrac{\partial C}{\partial x} + vC\right)_{x=0} = 0 \quad t > t_s$	$\left.\dfrac{\partial C}{\partial x}\right|_{x=L} = 0$	λC	$k_d\dfrac{\partial C}{\partial t}$	van Genuchten and Alves (1982, p. 66)
$C = C_i$	$\left(-D_x\dfrac{\partial C}{\partial x} + vC\right)_{x=0} = vC_0 \quad 0 < t \le t_s$ $\left(-D_x\dfrac{\partial C}{\partial x} + vC\right)_{x=0} = 0 \quad t > t_s$	$\left.\dfrac{dC}{dx}\right|_{x=\infty} = C(\infty,t) = 0$	λC	$k_d\dfrac{\partial C}{\partial t}$	van Genuchten and Alves (1982, p. 61)
$C = C_i$	$\left(-D_x\dfrac{\partial C}{\partial x} + vC\right)_{x=0} = vC_0 \quad 0 < t \le t_s$ $\left(-D_x\dfrac{\partial C}{\partial x} + vC\right)_{x=0} = 0 \quad t > t_s$	$\left.\dfrac{dC}{dx}\right|_{x=\infty} = C(\infty,t) = 0$	λC	$\alpha(k_d C - S)$	Toride et al. (1993)
$C = M\delta(x)$	$\left(-D_x\dfrac{\partial C}{\partial x} + vC\right)_{x=0} = 0$	$\left.\dfrac{\partial C}{\partial x}\right|_{x=\infty} = 0$	0	$\alpha(k_d C - S)$	Goltz (1986)
$C = C_i,\ 0 < x \le x_0$ $C = 0 \quad x > x_0$	$\left(-D_x\dfrac{\partial C}{\partial x} + vC\right)_{x=0} = 0$	$\left.\dfrac{\partial C}{\partial x}\right|_{x=\infty} = 0$	0	$\alpha(k_d C - S)$	Lindstrom and Narasimhan (1973)

References

Goltz, M.N., *Three-Dimensional Analytical Modeling of Diffusion-Limited Solute Transport*, Ph.D. Dissertation, Department of Civil Engineering, Stanford University, 1986.

Goltz, M.N. and P.V. Roberts, Three-dimensional solutions for solute transport in an infinite medium with mobile and immobile zones, *Water Resources Research*, 22 (7):1139–1148, 1986.

Lindstrom, F.T. and M.N. Narasimhan, Mathematical theory of a kinetic model for dispersion of previously distributed chemicals in a sorbing porous medium, *SIAM Journal on Applied Mathematics*, 24(4):496–510, 1973.

Toride, N., F. J. Leij, and M.Th. van Genuchten, A comprehensive set of analytical solutions for nonequilibrium solute transport with first-order decay and zero-order production, *Water Resources Research*, 29, 2167–2182, 1993.

van Genuchten, M. Th. and W.J. Alves, *Analytical Solutions of the One-Dimensional Convective–Dispersive Solute Transport Equation*, Agricultural Research Service Technical Bulletin 1661, 1982.

D

User Instructions for AnaModelTool Software

AnaModelTool is a user-friendly MATLAB®-based program that analytically solves the one-, two-, and three-dimensional Laplace transformed advection-dispersion-reaction equation for various initial and boundary conditions. The program then numerically inverts the Laplace-transformed solution into real-time and outputs concentration vs. time and concentration vs. space plots. Source code for AnaModelTool is provided as m-files and is also included in Appendix N.

To use AnaModelTool you must download the file AnaModelInstaller_mcr.exe to a computer running Windows from the website www.wiley.com/go/Goltz/solute_transport_in_groundwater.

To download the file from the site, click on "Software" in the Resources Tab. Follow the instructions to enter the password, and click on "Software" and "Installation file" t download AnaModelInstaller_mcr.zip. Unzip the file using WinZip and you will have the file: AnaModelInstaller_mcr.exe. Double-click to run AnaModelInstaller_mcr.exe. Windows may ask if you want to start an unrecognized app. Click on Yes, Okay, or Run Anyway (under the Windows "more info" tab) to permit installation. Both the MATLAB® runtime library (MCR) and the executable program AnaModelTool.exe will be installed on your computer. For the MATLAB® runtime library you will need to accept the license agreement. Note that the files being installed are large, so installation may take some time to complete—**be patient**. After installation, AnaModelTool.exe will be found as a Desktop icon, if you chose that option during installation. The executable file will also be in the default folder C:\Program Files\AnaModelTool\application. Double-click to run AnaModelTool.exe and the AnaModelTool home screen will appear.

In the upper left hand corner of the home screen is a drop down menu that allows the user to select the model to be solved.

Analytical Modeling of Solute Transport in Groundwater: Using Models to Understand the Effect of Natural Processes on Contaminant Fate and Transport, First Edition. Mark Goltz and Junqi Huang.
© 2017 John Wiley & Sons, Inc. Published 2017 by John Wiley & Sons, Inc.
Companion Website: www.wiley.com/go/Goltz/solute_transport_in_groundwater

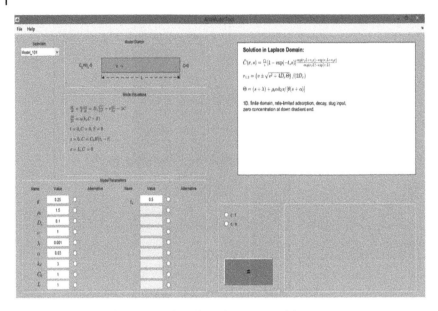

Figure D.1 AnaModelTool screen shot after selecting a model to run.

Figure D.1 is a typical AnaModelTool screen (in this case, Model 101 has been selected). Next to the model selection box is a sketch that shows the conceptual model along with the boundary conditions. The model equations (including the initial and boundary conditions) are shown below the sketch. In the upper right-hand corner is the Laplace domain solution, which will be numerically inverted by AnaModelTool.

Directly below the model equations is the model parameter input area, where the user can type in parameter values. Note units are not specified for the parameter values, though the values that are input must have consistent units. That is, if pore velocity (v) is input with units of m/d, then all parameters that have length or time units must also have units of m and d. The radio button allows the user to select two values for one of the parameters. The buttons above the equal sign allow the user to specify the desired output; in this case, concentration (c) versus time (t) or concentration (c) versus space (x). Once the desired output is chosen, the user gets to input the values of the independent variables.

Figure D.2 shows the screen after the equal sign button has been pressed. Note that in the lower right-hand corner, the independent variable values are indicated. In this case, the user selected a concentration versus time breakthrough curve for output, so a range of times were specified (0–2), calculated using an increment of 0.01, at a distance of 0.5 from the left-hand boundary ($x = 0$).

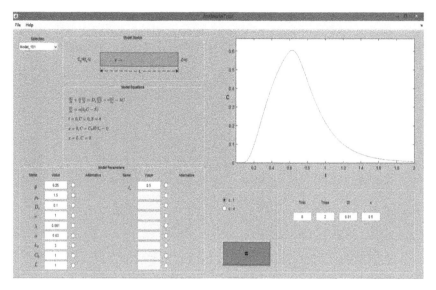

Figure D.2 AnaModelTool screen shot after choosing the independent parameter values and running model.

Also note in Figure D.2 that after the equal sign has been pressed, the Laplace domain solution has been replaced with a graphical output, in this case showing the breakthrough curve for the desired independent parameter values.

In the top upper left of the screen, choosing "file" lets the user save the raw data or export the plot.

E

Useful Laplace Transforms

$$\overline{C}(s) = \int_0^{\infty} e^{-st} C(t)\, dt$$

$C(t)$	$\overline{C}(s)$
a	$\dfrac{a}{s}$
at	$\dfrac{a}{s^2}$
$e^{-at} C(t)$	$\overline{C}(s+a)$
$aC_1(t)+bC_2(t)$	$a\overline{C}_1(s) + b\overline{C}_2(s)$
$\dfrac{dC(t)}{dt}$	$s\overline{C}(s) - C(0)$
$\dfrac{dC(t)}{dx}$	$\dfrac{d\overline{C}(s)}{dx}$
$\dfrac{d^2 C(t)}{dx^2}$	$\dfrac{d^2 \overline{C}(s)}{dx^2}$
Dirac delta function: $\delta(t - t_s)$	$e^{-t_s s}$
Heaviside step function: $H(t - t_s)$	$\dfrac{e^{-t_s s}}{s}$
Time pulse over interval $1 - H(t - t_s)$	$\dfrac{(1 - e^{-t_s s})}{s}$

E.1   Laplace Transforms from van Genuchten and Alves (1982)

The following abbreviations are used in the table:

$$A = \frac{1}{\sqrt{\pi t}} e^{\frac{-x^2}{4t}}$$

Analytical Modeling of Solute Transport in Groundwater: Using Models to Understand the Effect of Natural Processes on Contaminant Fate and Transport, First Edition. Mark Goltz and Junqi Huang.
© 2017 John Wiley & Sons, Inc. Published 2017 by John Wiley & Sons, Inc.
Companion Website: www.wiley.com/go/Goltz/solute_transport_in_groundwater

$$B = \text{erfc}\left(\frac{x}{2\sqrt{t}}\right)$$

$$C = e^{(a^2 t - ax)}\text{erfc}\left(\frac{x}{2\sqrt{t}} - a\sqrt{t}\right)$$

$$D = e^{(a^2 t + ax)}\text{erfc}\left(\frac{x}{2\sqrt{t}} + a\sqrt{t}\right)$$

$C(t)$	$\overline{C}(s)$
$\dfrac{x}{2t}A$	$e^{-x\sqrt{s}}$
A	$\dfrac{e^{-x\sqrt{s}}}{\sqrt{s}}$
B	$\dfrac{e^{-x\sqrt{s}}}{s}$
$2tA - xB$	$\dfrac{e^{-x\sqrt{s}}}{s\sqrt{s}}$
$\dfrac{1}{2}(x^2 + 2t)B - xtA$	$\dfrac{e^{-x\sqrt{s}}}{s^2}$
$(4t)^{\frac{n}{2}} i^n \text{erfc}\left(\dfrac{x}{\sqrt{t}}\right)$ $(n = 0, 1, 2, \ldots)$	$\dfrac{e^{-x\sqrt{s}}}{s^{1+n/2}}$
$A + \dfrac{a}{2}(C - D)$	$\dfrac{\sqrt{s}e^{-x\sqrt{s}}}{s - a^2}$
$\dfrac{1}{2}[C + D]$	$\dfrac{e^{-x\sqrt{s}}}{s - a^2}$
$\dfrac{1}{2a}[C - D]$	$\dfrac{e^{-x\sqrt{s}}}{\sqrt{s}(s - a^2)}$
$At + \dfrac{1}{4a}(1 - ax + 2a^2 t)C - \dfrac{1}{4a}(1 + ax + 2a^2 t)D$	$\dfrac{\sqrt{s}e^{-x\sqrt{s}}}{(s - a^2)^2}$
$\dfrac{1}{4a}(2at - x)C + \dfrac{1}{4a}(2at + x)D$	$\dfrac{e^{-x\sqrt{s}}}{(s - a^2)^2}$
$A\dfrac{t}{a^2} - \dfrac{1}{4a^3}(1 + ax - 2a^2 t)C + \dfrac{1}{4a^3}(1 - ax - 2a^2 t)D$	$\dfrac{e^{-x\sqrt{s}}}{\sqrt{s}(s - a^2)^2}$
$\left(\dfrac{x}{2t} - a\right)A + a^2 D$	$\dfrac{\sqrt{s}e^{-x\sqrt{s}}}{a + \sqrt{s}}$
$A - aD$	$\dfrac{e^{-x\sqrt{s}}}{a + \sqrt{s}}$
D	$\dfrac{e^{-x\sqrt{s}}}{\sqrt{s}(a + \sqrt{s})}$

$$\frac{1}{a}[B - D]$$

$$\frac{e^{-x\sqrt{s}}}{s(a + \sqrt{s})}$$

$$A\frac{2t}{a} - \frac{1}{a^2}(1 + ax)B + \frac{1}{a^2}D$$

$$\frac{e^{-x\sqrt{s}}}{s\sqrt{s}(a + \sqrt{s})}$$

$$\frac{1}{a^3}\left(1 + ax + a^2t + \frac{a^2x^2}{2}\right)B - \frac{1}{a^3}D - \frac{t}{a^2}(2 + ax)A$$

$$\frac{e^{-x\sqrt{s}}}{s^2(a + \sqrt{s})}$$

$$\frac{1}{(-a)^n}\left(D - \sum_{r=0}^{n-1}(-2a\sqrt{t})^r i^r \operatorname{erfc}\left(\frac{x}{2\sqrt{t}}\right)\right)$$

$$\frac{e^{-x\sqrt{s}}}{s^{n+1/2}(a + \sqrt{s})}$$

$$\frac{1}{4}C + \frac{1}{4}(3 + 2ax + 4a^2t)D - atA$$

$$\frac{\sqrt{s}e^{-x\sqrt{s}}}{(s - a^2)(a + \sqrt{s})}$$

$$\frac{1}{4a}C - \frac{1}{4a}(1 + 2ax + 4a^2t)D + At$$

$$\frac{e^{-x\sqrt{s}}}{(s - a^2)(a + \sqrt{s})}$$

$$\frac{1}{4a^2}C + \frac{1}{4a^2}(-1 + 2ax + 4a^2t)D - \frac{At}{a}$$

$$\frac{e^{-x\sqrt{s}}}{\sqrt{s}(s - a^2)(a + \sqrt{s})}$$

$$\frac{1}{4a^3}C + \frac{1}{4a^3}(3 - 2ax - 4a^2t)D + \frac{At}{a^2} - \frac{B}{a^3}$$

$$\frac{e^{-x\sqrt{s}}}{s(s - a^2)(a + \sqrt{s})}$$

$$\frac{t}{4a^2}(1 + ax + 2a^2t)A + \frac{1}{16a^3}(-1 - 2ax + 4a^2t)C$$
$$- \frac{D}{16a^3}(-1 + 2a^2(x + 2at)^2 + 4a^2t)$$

$$\frac{e^{-x\sqrt{s}}}{(s - a^2)^2(a + \sqrt{s})}$$

$$(1 + 2a^2t)A - a(2 + ax + 2a^2t)D$$

$$\frac{\sqrt{s}e^{-x\sqrt{s}}}{(a + \sqrt{s})^2}$$

$$-2atA + (1 + ax + 2a^2t)D$$

$$\frac{e^{-x\sqrt{s}}}{(a + \sqrt{s})^2}$$

$$2tA - (x + 2at)D$$

$$\frac{e^{-x\sqrt{s}}}{\sqrt{s}(a + \sqrt{s})^2}$$

$$\frac{1}{a^2}(-1 + ax + 2a^2t)D + \frac{1}{a^2}B - \frac{2t}{a}A$$

$$\frac{e^{-x\sqrt{s}}}{s(a + \sqrt{s})^2}$$

$$\frac{4t}{a^2}A - \frac{1}{a^3}(2 + ax)B - \frac{D}{a^3}(-2 + ax + 2a^2t)$$

$$\frac{e^{-x\sqrt{s}}}{s\sqrt{s}(a + \sqrt{s})^2}$$

$$\frac{1}{a^4}\left(3 + 2ax + a^2t + \frac{a^2x^2}{2}\right)B + \frac{1}{a^4}(-3 + ax + 2a^2t)D$$
$$- \frac{At}{a^3}(6 + ax)$$

$$\frac{e^{-x\sqrt{s}}}{s^2(a + \sqrt{s})^2}$$

$$\frac{t}{2}(3 + ax + 2a^2t)A + \frac{C}{8a} - \frac{1}{8a}(1 + 6ax + 16a^2t + 2a^2(x + 2at)^2)D$$

$$\frac{\sqrt{s}e^{-x\sqrt{s}}}{(s - a^2)(a + \sqrt{s})^2}$$

$$-\frac{t}{2a}(1 + ax + 2a^2t)A + \frac{C}{8a^2}$$
$$+ \frac{1}{8a^2}(-1 + 2ax + 8a^2t + 2a^2(x + 2at)^2)D$$

$$\frac{e^{-x\sqrt{s}}}{(s - a^2)(a + \sqrt{s})^2}$$

$$\frac{t}{2a^2}(-1+ax+2a^2t)A+\frac{1}{8a^3}C-\frac{D}{8a^3}(1-2ax+2a^2(x+2at)^2)$$
$$\frac{e^{-x\sqrt{s}}}{\sqrt{s}(s-a^2)(a+\sqrt{s})^2}$$

$$-\frac{t}{12a^3}(-3+4a^2t+a^2(x+2at)^2)A-\frac{1}{16a^4}(1+ax-2a^2t)C$$
$$+\frac{D}{16a^4}\left(1+a(4a^2t-1)(x+2at)+\frac{4}{3}a^3(x+2at)^3\right)$$
$$\frac{e^{-x\sqrt{s}}}{(s-a^2)^2(a+\sqrt{s})^2}$$

$$\left(1+2ax+5a^2t+\frac{a^2}{2}(x+2at)^2\right)D-at(4+ax+2a^2t)A$$
$$\frac{\sqrt{s}\,e^{-x\sqrt{s}}}{(a+\sqrt{s})^3}$$

$$-\left(x+3at+\frac{a}{2}(x+2at)^2\right)D+t(2+ax+2a^2t)A$$
$$\frac{e^{-x\sqrt{s}}}{(a+\sqrt{s})^3}$$

$$\left(t+\frac{1}{2}(x+2at)^2\right)D-t(x+2at)A$$
$$\frac{e^{-x\sqrt{s}}}{\sqrt{s}(a+\sqrt{s})^3}$$

$$\frac{t}{12a^2}(-3+3ax+14a^2t+2a^2(x+2at)^2)A+\frac{1}{16a^3}(C+(2ax-1)D)$$
$$-\frac{D}{24a}(3(x+2at)(x+6at)+2a(x+2at)^3)$$
$$\frac{e^{-x\sqrt{s}}}{(s-a^2)(a+\sqrt{s})^3}$$

$$t\left(2+2ax+\frac{16}{3}a^2t+\frac{a^2}{3}(x+2at)^2\right)A$$
$$-\left(x+at+a(x+2at)(x+3at)+\frac{a^2}{3}(x+2at)^3\right)D$$
$$\frac{\sqrt{s}\,e^{-x\sqrt{s}}}{(a+\sqrt{s})^4}$$

$$\left(t+\frac{1}{2}(x+2at)(x+4at)+\frac{a}{6}(x+2at)^3\right)D$$
$$-\frac{t}{3}(3x+10at+a(x+2at)^2)A$$
$$\frac{e^{-x\sqrt{s}}}{(a+\sqrt{s})^4}$$

$$\frac{t}{3}(4t+(x+2at)^2)A-\left(t(x+2at)+\frac{1}{6}(x+2at)^2\right)D$$
$$\frac{e^{-x\sqrt{s}}}{\sqrt{s}(a+\sqrt{s})^4}$$

$$\frac{C}{2(a+b)}-\frac{D}{2(a-b)}+\frac{b}{a^2-b^2}e^{(b^2t+bx)}\mathrm{erfc}\left(\frac{x}{2\sqrt{t}}+b\sqrt{t}\right)$$
$$\frac{e^{-x\sqrt{s}}}{(s-a^2)(b+\sqrt{s})}$$

Reference

van Genuchten, M.Th. and W.J. Alves, *Analytical Solutions of the One-Dimensional Convective–dispersive Solute Transport Equation,* Agricultural Research Service Technical Bulletin 1661, 1982.

F

Solution to Three-Dimensional ADR Equation with First-Order Degradation Kinetics for an Instantaneous Point Source Using Laplace and Fourier Transforms

In this appendix, we solve the three-dimensional ADR equation with first-order degradation kinetics, as formulated in Equation (2.21b) with $\frac{\partial C}{\partial t}\big|_{rxn} = \lambda C$. We assume the system is initially uncontaminated with a Dirac point source at $x = y = z = 0$, and first-type boundary conditions with $C = 0$ at $x = y = z = \infty$. Rewriting the ADR with its initial and boundary conditions (IC/BCs) gives

$$\frac{\partial C}{\partial t} = -\frac{v}{R}\frac{\partial C}{\partial x} + \frac{D_x}{R}\frac{\partial^2 C}{\partial x^2} + \frac{D_y}{R}\frac{\partial^2 C}{\partial y^2} + \frac{D_z}{R}\frac{\partial^2 C}{\partial z^2} - \frac{1}{R}\lambda C \tag{F.1}$$

$$C(x, y, z, t = 0) = \frac{M}{R}\delta(x)\delta(y)\delta(z) \tag{F.2a}$$

$$C(x = y = z = \infty, t) = 0 \tag{F.2b}$$

Our approach will be to use Laplace transforms in time and Fourier transforms in space to convert the PDE with its IC/BCs into an equation that can be straightforwardly solved, and then invert the Laplace and Fourier transformed solution into real time and space.

The Laplace transform is defined in Appendix A, and the Fourier transform ($\widehat{F}(p, y, z, t)$) is defined as

$$\widehat{F}(p, y, z, t) = \int_{-\infty}^{\infty} F(x, y, z, t)e^{-ipx}dx$$

where p is the Fourier transform variable in the x-direction. We can similarly define q and u as Fourier transform variables in the y- and z-directions, respectively.

A useful property of the Fourier transform is

$$\frac{\partial F(x, y, z, t)}{\partial x} = ip\widehat{F}(p, y, z, t) \text{ and therefore } \frac{\partial^2 F(x, y, z, t)}{\partial x^2} = -p^2\widehat{F}(p, y, z, t)$$

Also, we see that based on the definition of the Fourier transform and the Dirac delta function, $\widehat{F}(p, q, u, t) = M$ for $F(x, y, z, t) = M\delta(x)\delta(y)\delta(z)$

Analytical Modeling of Solute Transport in Groundwater: Using Models to Understand the Effect of Natural Processes on Contaminant Fate and Transport, First Edition. Mark Goltz and Junqi Huang.
© 2017 John Wiley & Sons, Inc. Published 2017 by John Wiley & Sons, Inc.
Companion Website: www.wiley.com/go/Goltz/solute_transport_in_groundwater

Defining $v' = \frac{v}{R}, D'_x = \frac{D_x}{R}, D'_y = \frac{D_y}{R}, D'_z = \frac{D_z}{R}, \lambda' = \frac{\lambda}{R}, M' = \frac{M}{R}$ we apply the Laplace transformation to Equation (A.1) and BC (A.2b) to obtain

$$s\overline{C} - C(x, y, z, 0) = -v'\frac{d\overline{C}}{dx} + D'_x\frac{d^2\overline{C}}{dx^2} + D'_y\frac{d^2\overline{C}}{dy^2}$$

$$+ D'_z\frac{d^2\overline{C}}{dz^2} - \lambda'\overline{C} \tag{F.3}$$

$$\overline{C}(x = y = z = \infty, s) = 0 \tag{F.4}$$

Substituting in the IC (F.2a) and rearranging, we have

$$-v'\frac{d\overline{C}}{dx} + D'_x\frac{d^2\overline{C}}{dx^2} + D'_y\frac{d^2\overline{C}}{dy^2} + D'_z\frac{d^2\overline{C}}{dz^2} - (\lambda' + s)\overline{C}$$

$$= -M'\delta(x)\delta(y)\delta(z) \tag{F.5}$$

$$\overline{C}(x = y = z = \infty, s) = 0 \tag{F.6}$$

Now we take the Fourier transform in the x-, y-, and z-directions to find

$$-v'ip\hat{\overline{C}} - D'_xp^2\hat{\overline{C}} - D'_yq^2\hat{\overline{C}} - D'_zu^2\hat{\overline{C}} - (\lambda' + s)\hat{\overline{C}} = -M'$$

where $\hat{\overline{C}}$ is the Laplace and Fourier transformed function. Rearranging

$$\hat{\overline{C}}(p, q, u, s) = \frac{M'}{D'_xp^2 + D'_yq^2 + D'_zu^2 + v'ip + (\lambda' + s)} \tag{F.7}$$

It is now necessary to sequentially invert the Fourier solution into x-, y-, and z-space. The formula to invert the Fourier transformed solution into x-space is

$$F(x, y, z, t) = \frac{1}{2\pi}\int_{-\infty}^{\infty} \hat{F}(p, y, z, t)e^{ipx}dp \tag{F.8}$$

Analogous expressions are used to invert the Fourier transformed solution into y- and z-space. Details for inverting the Fourier solution into x-, y-, and z-space are shown in Goltz and Roberts (1986). The inverted Fourier solution, still in Laplace time, is

$$\overline{C}(x, y, z, s) = \frac{F}{G}\exp\left[-G\sqrt{\frac{v'^2}{4D'_x} + (s + \lambda')}\right] \tag{F.9}$$

where

$$F = \frac{M'\exp\left[\frac{v'x}{2D'_x}\right]}{4\pi\sqrt{D'_xD'_yD'_z}}$$

and

$$G = \sqrt{\frac{x^2}{D_x'} + \frac{y^2}{D_y'} + \frac{z^2}{D_z'}}$$

We then use Laplace inversion (Equation (A.8)), following standard Laplace inversion methods (e.g., Polyanin and Manzhirov, 2007) to find the solution in real time and space:

$$C(x, y, z, t) = \frac{M' \exp\left[-\frac{(x-v't)^2}{4D_x't} - \frac{y^2}{4D_y't} - \frac{z^2}{4D_z't} - \lambda't\right]}{(4\pi t)^{3/2}\sqrt{D_x'D_y'D_z'}} \tag{F.10}$$

References

Goltz, M.N. and P.V. Roberts, Three-dimensional solutions for solute transport in an infinite medium with mobile and immobile zones, *Water Resources Research*, 22(7): 1139–1148, 1986.

Polyanin, A.D. and A.V. Manzhirov, *Handbook of Mathematics for Engineers and Scientists*, pp. 1173, Chapman & Hall/CRC, Boca Raton, 2007.

G

Solution to Three-Dimensional ADR Equation with Zeroth-Order Degradation Kinetics for an Instantaneous Point Source Using Laplace and Fourier Transforms

In this appendix, we solve the three-dimensional ADR equation with zeroth-order degradation kinetics, as formulated in Equation (2.21b) with $\frac{\partial C}{\partial t}\big|_{rxn} = -k_0$. We assume the system is initially uncontaminated with a Dirac point source at $x = y = z = 0$, and first-type boundary conditions with $C = 0$ at $x = y = z = \infty$. Rewriting the ADR with its initial and boundary conditions (IC/BCs) gives

$$\frac{\partial C}{\partial t} = -\frac{v}{R}\frac{\partial C}{\partial x} + \frac{D_x}{R}\frac{\partial^2 C}{\partial x^2} + \frac{D_y}{R}\frac{\partial^2 C}{\partial y^2} + \frac{D_z}{R}\frac{\partial^2 C}{\partial z^2} - \frac{k_0}{R} \tag{G.1}$$

$$C(x, y, z, t = 0) = \frac{M}{R}\delta(x)\delta(y)\delta(z) \tag{G.2a}$$

$$C(x = y = z = \infty, t) = 0 \tag{G.2b}$$

This PDE with IC/BCs can be solved in a number of different ways. In this appendix, we follow the approach used in Appendix F by taking Laplace transforms in time and Fourier transforms in space.

Defining $v' = \frac{v}{R}, D'_x = \frac{D_x}{R}, D'_y = \frac{D_y}{R}, D'_z = \frac{D_z}{R}, k_0' = \frac{k_0}{R}, M' = \frac{M}{R}$ we apply the Laplace transformation to Equation G.1 and BC G.2b to obtain

$$s\overline{C} - C(x, y, z, 0) = -v'\frac{d\overline{C}}{dx} + D'_x\frac{d^2\overline{C}}{dx^2} + D'_y\frac{d^2\overline{C}}{dy^2} + D'_z\frac{d^2\overline{C}}{dz^2} - \frac{k_0'}{s} \tag{G.3}$$

$$\overline{C}(x = y = z = \infty, s) = 0 \tag{G.4}$$

Substituting in the IC G.2a and rearranging, we have

$$-v'\frac{d\overline{C}}{dx} + D'_x\frac{d^2\overline{C}}{dx^2} + D'_y\frac{d^2\overline{C}}{dy^2} + D'_z\frac{d^2\overline{C}}{dz^2} - s\overline{C} - \frac{k_0'}{s} = -M'\delta(x)\delta(y)\delta(z) \tag{G.5}$$

$$\overline{C}(x = y = z = \infty, s) = 0 \tag{G.6}$$

Analytical Modeling of Solute Transport in Groundwater: Using Models to Understand the Effect of Natural Processes on Contaminant Fate and Transport, First Edition. Mark Goltz and Junqi Huang.
© 2017 John Wiley & Sons, Inc. Published 2017 by John Wiley & Sons, Inc.
Companion Website: www.wiley.com/go/Goltz/solute_transport_in_groundwater

Now we take the Fourier transform in the x-, y-, and z-directions to find

$$-v'ip\hat{\bar{C}} - D_x'p^2\hat{\bar{C}} - D_y'q^2\hat{\bar{C}} - D_z'u^2\hat{\bar{C}} - s\hat{\bar{C}} - \frac{k_0'}{s}(2\pi)^3\delta(p)\delta(q)\delta(u) = -M'$$

where $\hat{\bar{C}}$ is the Laplace and Fourier transformed function. Rearranging

$$\hat{\bar{C}}(p,q,u,s) = \frac{M' - \frac{k_0'}{s}(2\pi)^3\delta(p)\delta(q)\delta(u)}{D_x'p^2 + D_y'q^2 + D_z'u^2 + v'ip + s} \tag{G.7}$$

It is now necessary to sequentially invert the Fourier solution into x-, y-, and z-space. The formula to invert the Fourier transformed solution into x-space is

$$F(x,y,z,t) = \frac{1}{2\pi}\int_{-\infty}^{\infty}\hat{F}(p,y,z,t)e^{ipx}\,dp \tag{G.8}$$

Analogous expressions are used to invert the Fourier transformed solution into y- and z-space. Details for inverting the Fourier solution into x-, y-, and z-space are shown in Goltz and Roberts (1986). The inverted Fourier solution, still in Laplace time, is

$$\bar{C}(x,y,z,s) = \frac{F}{G}\exp\left[-G\sqrt{\frac{v'^2}{4D_x'} + s}\right] - \frac{k_0'}{s^2} \tag{G.9}$$

where

$$F = \frac{M'\exp\left[\dfrac{v'x}{2D_x'}\right]}{4\pi\sqrt{D_x'D_y'D_z'}}$$

and

$$G = \sqrt{\frac{x^2}{D_x'} + \frac{y^2}{D_y'} + \frac{z^2}{D_z'}}$$

We then use Laplace inversion (Equation (A.8)), following standard Laplace inversion methods (e.g., Polyanin and Manzhirov, 2007) to find the solution in real time and space:

$$C(x,y,z,t) = \frac{M'\exp\left[-\dfrac{(x-v't)^2}{4D_x't} - \dfrac{y^2}{4D_y't} - \dfrac{z^2}{4D_z't}\right]}{(4\pi t)^{3/2}\sqrt{D_x'D_y'D_z'}} - k_0't \tag{G.10}$$

References

Goltz, M.N. and P.V. Roberts, Three-dimensional solutions for solute transport in an infinite medium with mobile and immobile zones, *Water Resources Research*, 22(7): 1139–1148, 1986.

Polyanin, A.D. and A.V. Manzhirov, *Handbook of Mathematics for Engineers and Scientists*, pp. 1173, Chapman & Hall/CRC, Boca Raton, 2007.

H

Solutions to the Three-Dimensional ADR in Literature

Analytical Modeling of Solute Transport in Groundwater: Using Models to Understand the Effect of Natural Processes on Contaminant Fate and Transport, First Edition. Mark Goltz and Junqi Huang.
© 2017 John Wiley & Sons, Inc. Published 2017 by John Wiley & Sons, Inc.
Companion Website: www.wiley.com/go/Goltz/solute_transport_in_groundwater

$$\frac{\partial C}{\partial t} = -v\frac{\partial C}{\partial x} + D_x\frac{\partial^2 C}{\partial x^2} + D_y\frac{\partial^2 C}{\partial y^2} + D_z\frac{\partial^2 C}{\partial z^2} - \frac{\partial C}{\partial t}\bigg|_{rxn} - \frac{\rho_b}{\theta}\frac{\partial S}{\partial t}$$

Initial condition ($t=0$)	BC x	BC y	BC z	$\dfrac{\partial C}{\partial t}\bigg|_{rxn}$	$\dfrac{\partial S}{\partial t}$	Reference
$C = M\delta(x)\delta(y)\delta(z)$*	$C = \dfrac{\partial C}{\partial x} = 0$ $x = \pm\infty$	$C = \dfrac{\partial C}{\partial y} = 0$ $y = \pm\infty$	$C = \dfrac{\partial C}{\partial z} = 0$ $z = \pm\infty$	λC	$k_d\dfrac{\partial C}{\partial t}$	Wexler (1992, p. 47)
$C = 0$ $0 < x < \infty$ $0 < y < w$ $0 < z < h$	$C(0, y_b \le y \le y_t, z_b \le z \le z_t, t) = C_0$ $C(0, y_b > y > y_t, z_b > z > z_t, t) = 0$ $C(\infty, y, z, t) = \dfrac{\partial C(\infty, y, z, t)}{\partial x} = 0$	$\dfrac{\partial C}{\partial y} = 0$ for $y = 0$ and $y = w$	$\dfrac{\partial C}{\partial z} = 0$ for $z = 0$ and $z = h$	λC	$k_d\dfrac{\partial C}{\partial t}$	Wexler (1992, p. 51)
$C = f(x,y,z)$ $0 < x < \infty$ $-\infty < y < \infty$ $-\infty < z < \infty$	$C(0,y,z,t) = g(y,z,t)$ $\dfrac{\partial C(\infty,y,z,t)}{\partial x} = 0$	$\dfrac{\partial C}{\partial y} = 0$ $y = \pm\infty$	$\dfrac{\partial C}{\partial z} = 0$ $z = \pm\infty$	λC and/or k_0	$k_d\dfrac{\partial C}{\partial t}$	Leij et al. (1991)
$C = f(x,y,z)$ $0 < x < \infty$ $-\infty < y < \infty$ $-\infty < z < \infty$	$-D_x\dfrac{\partial C}{\partial x} + vC = vC_0$, $x = 0$ $\dfrac{\partial C}{\partial x} = 0$ $x = \infty$	$\dfrac{\partial C}{\partial y} = 0$ $y = \pm\infty$	$\dfrac{\partial C}{\partial z} = 0$ $z = \pm\infty$	λC and/or k_0	$k_d\dfrac{\partial C}{\partial t}$	Leij et al. (1991)
$C = M\delta(x)\delta(y)\delta(z)$*	$C = 0$ $x = \pm\infty$	$C = 0$ $y = \pm\infty$	$C = 0$ $z = \pm\infty$	λC	$\alpha(k_d C - S)$	Carnahan and Remer (1984)

*Superposition can be applied to derive solutions for various 3-D source zone geometries in an infinite medium.

AnaModelTool Three-Dimensional Models (PDE and IC/BC)*

$$\frac{\partial C}{\partial t} = -v\frac{\partial C}{\partial x} + D_x\frac{\partial^2 C}{\partial x^2} + D_y\frac{\partial^2 C}{\partial y^2} + D_z\frac{\partial^2 C}{\partial z^2} - \lambda C - \frac{\rho_b}{\theta}\frac{\partial S}{\partial t}$$

$$\frac{\partial S}{\partial t} = \alpha(k_d C - S)$$

Initial condition (t = 0)	Boundary condition x	Boundary condition y	Boundary condition z	AnaModelTool model number
$C = 0$	$C(0, y_b \leq y \leq y_t, z_b \leq z \leq z_t, t) = C_0 H(t_s - t)$ $C(0, y_b > y > y_t, z_b > z > z_t, t) = 0$ $C(l, y, z, t) = 0$	$C(x, 0, z, t) = 0$ $C(x, w, z, t) = 0$	$C(x, y, 0, t) = 0$ $C(x, y, h, t) = 0$	301
$C = 0$	$C(0, y_b \leq y \leq y_t, z_b \leq z \leq z_t, t) = C_0 H(t_s - t)$ $C(0, y_b > y > y_t, z_b > z > z_t, t) = 0$ $C(l, y, z, t) = 0$	$\frac{\partial C}{\partial y}(x, 0, z, t) = 0$ $\frac{\partial C}{\partial y}(x, w, z, t) = 0$	$\frac{\partial C}{\partial z}(x, y, 0, t) = 0$ $\frac{\partial C}{\partial z}(x, y, h, t) = 0$	302
$C = 0$	$C(0, y_b \leq y \leq y_t, z_b \leq z \leq z_t, t) = C_0 H(t_s - t)$ $C(0, y_b > y > y_t, z_b > z > z_t, t) = 0$ $\frac{\partial C}{\partial x}(l, y, z, t) = 0$	$\frac{\partial C}{\partial y}(x, 0, z, t) = 0$ $\frac{\partial C}{\partial y}(x, w, z, t) = 0$	$\frac{\partial C}{\partial z}(x, y, 0, t) = 0$ $\frac{\partial C}{\partial z}(x, y, h, t) = 0$	303
$C = 0$	$C(0, y_b \leq y \leq y_t, z_b \leq z \leq z_t, t) = C_0 H(t_s - t)$ $C(0, y_b > y > y_t, z_b > z > z_t, t) = 0$ $C(\infty, y, z, t) = 0$	$C(x, 0, z, t) = 0$ $C(x, w, z, t) = 0$	$C(x, y, 0, t) = 0$ $C(x, y, h, t) = 0$	304
$C = 0$	$C(0, y_b \leq y \leq y_t, z_b \leq z \leq z_t, t) = C_0 H(t_s - t)$ $C(0, y_b > y > y_t, z_b > z > z_t, t) = 0$ $C(\infty, y, z, t) = 0$	$\frac{\partial C}{\partial y}(x, 0, z, t) = 0$ $\frac{\partial C}{\partial y}(x, w, z, t) = 0$	$\frac{\partial C}{\partial z}(x, y, 0, t) = 0$ $\frac{\partial C}{\partial z}(x, y, h, t) = 0$	305
$C = M\delta(x, y, z)$	$C(\pm\infty, y, z, t) = 0$	$C(x, \pm\infty, z, t) = 0$	$C(x, y, \pm\infty, t) = 0$	306

*Definitions of l, w, h, y_b, y_t, z_b, and z_t shown in Figure H.1.

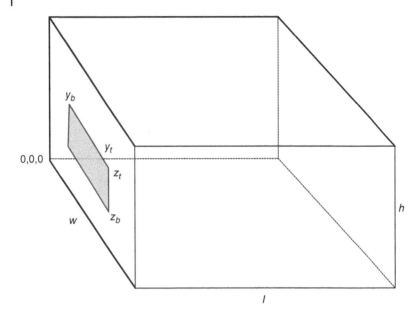

Figure H.1 Coordinates for three-dimensional aquifer system.

References

Carnahan, C.L. and J.S. Remer, Nonequilibrium and equilibrium sorption with a linear sorption isotherm during mass transport through an infinite porous medium: some analytical solutions, *Journal of Hydrology, 73*, 227–258, 1984.

Leij, F.J., T.H. Skaggs, and M.Th. van Genuchten, Analytical solutions for solute transport in three-dimensional semi-infinite porous media, *Water Resources Research, 27*(10): 2719–2733, 1991.

Wexler, E.J., *Analytical Solutions for One-, Two-, and Three-Dimensional Solute Transport in Ground-Water Systems with Uniform Flow*, Techniques of Water-Resources Investigations of the United States Geological Survey, Chapter B7, US Government Printing Office, 1992.

I

Derivation of the Long-Time First-Order Rate Constant to Model Decrease in Concentrations at a Monitoring Well Due to Advection, Dispersion, Equilibrium Sorption, and First-Order Degradation (Three-Dimensional Infinite System with an Instantaneous Point Source)

In this appendix, we calculate the first-order rate constant for the decrease in concentrations at long times due to advection, dispersion, equilibrium sorption, and first-order degradation. For the system defined in Appendix F (Dirac pulse in a three-dimensional infinite aquifer), the concentration as a function of space and time is shown in Equation (F.10):

$$C(x, y, z, t) = \frac{M' \exp\left[-\dfrac{(x - v't)^2}{4D'_x t} - \dfrac{y^2}{4D'_y t} - \dfrac{z^2}{4D'_z t} - \lambda' t\right]}{(4\pi t)^{3/2} \sqrt{D'_x D'_y D'_z}}$$

Rewriting this equation, using the definitions in Appendix F:

$$
\begin{aligned}
&C(x, y, z, t) \\
&= \frac{M \exp[vx/(2D_x)] \sqrt{R}}{(4\pi t)^{3/2} \sqrt{D_x D_y D_z}} \exp\left[-\frac{\left(\dfrac{x^2}{D_x} + \dfrac{y^2}{D_y} + \dfrac{z^2}{D_z}\right) R}{4t} - \left(\frac{v^2}{4D_x} + \lambda\right)\frac{t}{R}\right]
\end{aligned}
$$

$$(I.1)$$

The derivative with respect to time of Equation (I.1) is

$$
\begin{aligned}
\frac{\partial C}{\partial t} = &\frac{M \exp[vx/(2D_x)] \sqrt{R}}{8\pi^{3/2} t^{3/2} \sqrt{D_x D_y D_z}} \exp\left[-\frac{\left(\dfrac{x^2}{D_x} + \dfrac{y^2}{D_y} + \dfrac{z^2}{D_z}\right) R}{4t} - \left(\frac{v^2}{4D_x} + \lambda\right)\frac{t}{R}\right] \\
&\times \left[-\frac{3}{2} t^{-1} + \frac{\left(\dfrac{x^2}{D_x} + \dfrac{y^2}{D_y} + \dfrac{z^2}{D_z}\right) R}{4} t^{-2} - \left(\frac{v^2}{4D_x} + \lambda\right)\frac{1}{R}\right]
\end{aligned}
$$

$$(I.2)$$

Analytical Modeling of Solute Transport in Groundwater: Using Models to Understand the Effect of Natural Processes on Contaminant Fate and Transport, First Edition. Mark Goltz and Junqi Huang.
© 2017 John Wiley & Sons, Inc. Published 2017 by John Wiley & Sons, Inc.
Companion Website: www.wiley.com/go/Goltz/solute_transport_in_groundwater

By the definition of a first-order rate constant (k):

$$k = \frac{1}{C}\frac{\partial C}{\partial t}$$

we find

$$k = \left[-\frac{3}{2}t^{-1} + \frac{\left(\dfrac{x^2}{D_x} + \dfrac{y^2}{D_y} + \dfrac{z^2}{D_z}\right)R}{4}t^{-2} - \left(\frac{v^2}{4D_x} + \lambda\right)\frac{1}{R} \right] \tag{I.3}$$

and at long time, when

$$\left| -\frac{3}{2}t^{-1} + \frac{\left(\dfrac{x^2}{D_x} + \dfrac{y^2}{D_y} + \dfrac{z^2}{D_z}\right)R}{4}t^{-2} \right| \ll \left| \left(\frac{v^2}{4D_x} + \lambda\right)\frac{1}{R} \right|$$

$$k = -\left(\frac{v^2}{4D_x} + \lambda\right)\frac{1}{R} \tag{I.4}$$

J

Application of Aris' Method of Moments to Calculate Temporal Moments

Following the methodology of Appendix A, we convert Equation (5.6) and initial/boundary conditions (5.7) to Laplace time to obtain the following ODE with BCs:

$$s\overline{C}(x,s) - C(x, t = 0) = -v'\frac{d\overline{C}}{dx} + D'\frac{d^2\overline{C}}{dx^2} - \lambda'\overline{C} \tag{J.1}$$

$$\overline{C}(x = 0, s) = \frac{1 - e^{-t_s s}}{s} C_0 \tag{J.2a}$$

$$\frac{d\overline{C}}{dx}(x = \infty, s) = 0 \tag{J.2b}$$

where

$$v' = \frac{v}{R}, D' = \frac{D_x}{R}, \lambda' = \frac{\lambda}{R},$$

s is the Laplace transform variable and $\overline{C}(x, s)$ is the Laplace transform of $C(x, t)$.

Following Appendix A, we solve ODE (J.1) with BCs (J.2) to find the Laplace time solution:

$$\overline{C}(x,s) = \frac{C_0(1 - e^{-t_s s})}{s} e^{\left(\frac{v'}{2D'} - \frac{1}{2}\sqrt{\left(\frac{v'}{D'}\right)^2 + \frac{4(s+\lambda')}{D'}}\right)x} \tag{J.3}$$

Equations (5.2), (5.4), and (5.5) are then applied to obtain the temporal moments:

$$m_{0,t} = C_0 t_s e^{\frac{x(v'-a')}{2D'}} \tag{J.4}$$

$$\mu'_{1,t} = \frac{m_{1,t}}{m_{0,t}} = \frac{t_s}{2} + \frac{x}{a'} \tag{J.5}$$

Analytical Modeling of Solute Transport in Groundwater: Using Models to Understand the Effect of Natural Processes on Contaminant Fate and Transport, First Edition. Mark Goltz and Junqi Huang.
© 2017 John Wiley & Sons, Inc. Published 2017 by John Wiley & Sons, Inc.
Companion Website: www.wiley.com/go/Goltz/solute_transport_in_groundwater

$$\mu'_{2,t} = \frac{m_{2,t}}{m_{0,t}} = \frac{t_s^2(4D'\lambda'a' + v'^2a') + t_s(3v'^2x + 12D'\lambda'x) + 3x^2a' + 6D'x}{3a'^3}$$

(J.6)

$$\mu_{2,t} = \mu'_{2,t} - (\mu'_{1,t})^2 = \frac{t_s^2}{12} + \frac{2D'x}{a'^3}$$

(J.7)

where $a' = \sqrt{v'^2 + 4D'\lambda'}$.

K

Application of Modified Aris' Method of Moments to Calculate Spatial Moments Assuming Equilibrium Sorption

We convert Equation (5.28) into Fourier space to obtain the following ODE with IC:

$$\frac{d\overline{F}}{dt} = D'i^2p^2\overline{F} - v'ip\overline{F} - \lambda'\overline{F} = -(D'p^2 + v'ip + \lambda')\overline{F} \tag{K.1}$$

$$\overline{F}(p,0) = M' \tag{K.1a}$$

where

$$v' = \frac{v}{R}, D' = \frac{D_x}{R}, \lambda' = \frac{\lambda}{R}, M' = \frac{M}{R}$$

p is the Fourier transform variable and $\overline{F}(p,t)$ is the Fourier transform of $C(x,t)$. The solution of this first-order ODE in Fourier space is straightforward:

$$\overline{F}(p,t) = M'e^{-(D'p^2+v'ip+\lambda')t} \tag{K.2}$$

Equation (5.26) is then applied to obtain the spatial moments:

$$m_0 = M'e^{-\lambda't} \tag{K.3}$$

$$m_1 = M'e^{-\lambda't}v't \tag{K.4}$$

$$m_2 = M'e^{-\lambda't}[(v')^2t^2 + 2D't] \tag{K.5}$$

Alternatively, we can take the Laplace transform of Equation (K.1) to obtain

$$s\hat{\overline{C}}(p,s) - M' = D'i^2p^2\hat{\overline{C}} - v'ip\hat{\overline{C}} - \lambda'\hat{\overline{C}} = -(D'p^2 + v'ip + \lambda')\hat{\overline{C}}$$

where $\hat{\overline{C}}(p,s)$ is the Laplace and Fourier transform of $C(x,t)$.

Solving for $\hat{\overline{C}}(p,s)$ we find

$$\hat{\overline{C}} = \frac{M'}{D'p^2 + v'ip + \lambda' + s} \tag{K.6}$$

Taking the Laplace inversion of Equation (K.6) also results in Equation (K.2).

$$\overline{F}(p,t) = M'e^{-(D'p^2+v'ip+\lambda')t} \tag{K.2}$$

Analytical Modeling of Solute Transport in Groundwater: Using Models to Understand the Effect of Natural Processes on Contaminant Fate and Transport, First Edition. Mark Goltz and Junqi Huang.
© 2017 John Wiley & Sons, Inc. Published 2017 by John Wiley & Sons, Inc.
Companion Website: www.wiley.com/go/Goltz/solute_transport_in_groundwater

L

Application of Modified Aris' Method of Moments to Calculate Spatial Moments Assuming Rate-Limited Sorption

After Laplace and Fourier transform of Equation (5.32), we obtain

$$\hat{\bar{C}}(p,s) = \frac{M(s+\alpha)}{(D_xp^2 + ivp + s + \lambda + \alpha\beta)(s+\alpha) - \alpha^2\beta}$$

$$= \frac{M(s+\alpha)}{s^2 + (D_xp^2 + ivp + \lambda + \alpha + \alpha\beta)s + \alpha(D_xp^2 + ivp + \lambda)} \tag{L.1}$$

where s is the Laplace transform variable, p is the Fourier transform variable, $\hat{\bar{C}}(p,s)$ is the Laplace and Fourier transform of $C(x,t)$, and $\beta = \frac{\rho_b k_d}{\theta}$.

The zeros of the quadratic polynomial in the denominator of Equation (L.1) are

$$s_{1,2} = -\frac{1}{2}B \pm \frac{1}{2}q$$

where

$$q = \sqrt{B^2 - 4F}$$

$$B = D_xp^2 + ivp + \lambda + \alpha + \alpha\beta$$

$$F = \alpha(D_xp^2 + ivp + \lambda)$$

We obtain the inverse Laplace transform of (L.1) by taking the inverse transform of a rational polynomial. We thereby find the solution to Equation (5.32) in Fourier space:

$$\hat{C}(p,t) = \frac{M}{q}[(s_1 + \alpha)e^{s_1 t} - (s_2 + \alpha)e^{s_2 t}] \tag{L.2}$$

Analytical Modeling of Solute Transport in Groundwater: Using Models to Understand the Effect of Natural Processes on Contaminant Fate and Transport, First Edition. Mark Goltz and Junqi Huang.
© 2017 John Wiley & Sons, Inc. Published 2017 by John Wiley & Sons, Inc.
Companion Website: www.wiley.com/go/Goltz/solute_transport_in_groundwater

L.1 Zeroth Spatial Moment

Letting $p \rightarrow 0$:

$$m_0(t) = \left. \widehat{C} \right|_{p=0} = M[\Delta_1 e^{s_{10}t} - \Delta_2 e^{s_{20}t}] \qquad \text{(L.3)}$$

where

$$\Delta_1 = -\frac{1}{2}(\gamma - 1)$$

$$\Delta_2 = -\frac{1}{2}(\gamma + 1)$$

$$s_{10} = -\frac{1}{2}(\lambda + \alpha + \alpha\beta) + \frac{1}{2}q_0$$

$$s_{20} = -\frac{1}{2}(\lambda + \alpha + \alpha\beta) - \frac{1}{2}q_0$$

$$\gamma = \frac{1}{q_0}[\lambda + \alpha(\beta - 1)]$$

$$q_0 = \sqrt{(\lambda + \alpha + \alpha\beta)^2 - 4\alpha\lambda}$$

L.2 First Spatial Moment

Taking the first derivative of Equation (L.2) with respect to p and letting $p \rightarrow 0$:

$$m_1(t) = i\left.\frac{\partial \widehat{C}}{\partial p}\right|_{p=0} = M(\Omega_1 e^{s_{10}t} - \Omega_2 e^{s_{20}t}) \qquad \text{(L.4)}$$

where

$$\Omega_1 = \frac{1}{2q_0}v(1 - \gamma^2) + \frac{1}{4}v(\gamma - 1)^2 t$$

$$\Omega_2 = \frac{1}{2q_0}v(1 - \gamma^2) + \frac{1}{4}v(\gamma + 1)^2 t$$

L.3 Second Spatial Moment

Taking the second derivative of Equation (L.2) with respect to p and letting $p \rightarrow 0$:

$$m_2(t) = i^2 \left.\frac{\partial^2 \widehat{C}}{\partial p^2}\right|_{p=0} = -M(\Psi_1 e^{s_{10}t} - \Psi_2 e^{s_{20}t}) \qquad \text{(L.5)}$$

where

$$\Psi = 1 - \frac{1}{q_0}(1 - \gamma^2)\left[D_x + \frac{3}{2q_0}v_x^2\gamma\right] - \frac{1}{2}(1 - \gamma)^2\left[D_x + \frac{3}{2q_0}v^2(1 + \gamma)\right]t$$

$$+ \frac{1}{8}v^2(\gamma - 1)^3 t^2$$

$$\Psi = 2 - \frac{1}{q_0}(1 - \gamma^2)\left[D_x + \frac{3}{2q_0}v^2\gamma\right] + \frac{1}{2}(1 + \gamma)^2\left[D_x - \frac{3}{2q_0}v^2(1 - \gamma)\right]t$$

$$+ \frac{1}{8}v^2(\gamma - 1)^3 t^2$$

M

Derivation of Laplace Domain Solutions to a Model Describing Radial Advective/Dispersive/Sorptive Transport to an Extraction Well

The following equations describe advective/dispersive/sorptive transport of a chemical in a one-dimensional system, where sorption is modeled as a first-order process (Equations (2.22a) and (2.22b)).

$$\frac{\partial C}{\partial t} = -v\frac{\partial C}{\partial x} + D_x\frac{\partial^2 C}{\partial x^2} - \frac{\rho_b}{\theta}\frac{\partial S}{\partial t}$$

$$\frac{\partial S}{\partial t} = \alpha(k_d C - S)$$

For radial flow toward an extraction well in a confined aquifer of thickness b, located at radial coordinate $r = 0$, pumping at flow Q_w, where the dispersion coefficient is modeled using Equation (2.9), Equation (2.22a) and (2.22c) may be written as follows:

$$\frac{\partial C}{\partial t} = -v_r\frac{\partial C}{\partial r} + \alpha_r|v_r|\frac{\partial^2 C}{\partial r^2} - \frac{\rho_b}{\theta}\frac{\partial S}{\partial t} \tag{M.1a}$$

$$\frac{\partial S}{\partial t} = \alpha(k_d C - S) \tag{M.1b}$$

$$v_r = \frac{-Q_w}{2\pi r b\theta} \tag{M.1c}$$

where α_r is the dispersivity in the radial direction.

For an extraction well of radius r_w at the center of a contaminated zone of radius r_b in an infinite confined aquifer, the following initial and boundary conditions apply:

$$C(r, t = 0) = C_0, \quad r_w \leq r \leq r_b$$
$$S(r, t = 0) = k_d C_0, \quad r_w \leq r \leq r_b$$
$$C(r, t = 0) = S(r, t = 0) = 0, \quad r > r_b \tag{M.2a}$$

$$v_r C(r = r_b, t) + \alpha_r v_r \frac{\partial C(r = r_b, t)}{\partial r} = 0 \tag{M.2b}$$

Analytical Modeling of Solute Transport in Groundwater: Using Models to Understand the Effect of Natural Processes on Contaminant Fate and Transport, First Edition. Mark Goltz and Junqi Huang.
© 2017 John Wiley & Sons, Inc. Published 2017 by John Wiley & Sons, Inc.
Companion Website: www.wiley.com/go/Goltz/solute_transport_in_groundwater

$$\frac{\partial C(r = r_w, t)}{\partial r} = 0 \tag{M.2c}$$

Equations (M.2a) simply state the initial condition, which assumes dissolved contaminant is at a uniform concentration C_0 within a circular area of radius r_b, and sorbed contaminant is at equilibrium with the dissolved contaminant. Outside the contaminated zone, contaminant concentrations are set to zero. Equation (M.2b) states that the total mass flux inward at the outer boundary $(r = r_b)$ must always be zero, since initially, there is no contaminant mass at $r > r_b$. Equation (M.2c) states the boundary condition at the well radius and is based on the assumption that at any time, the concentration inside the well bore is equal to that entering the well from surrounding media (Chen and Woodside, 1988). This implies a zero concentration gradient at the interface between the well and its immediate adjacent aquifer.

Following the procedures in Section 2.5, we can nondimensionalize Equations (M.1) and initial/boundary conditions (M.2) to obtain

$$\frac{\partial \tilde{C}}{\partial T} = \frac{1}{X}\frac{\partial \tilde{C}}{\partial X} + \frac{1}{X}\frac{\partial^2 \tilde{C}}{\partial X^2} - \beta\frac{\partial \tilde{S}}{\partial T} \tag{M.3a}$$

$$\frac{\partial \tilde{S}}{\partial T} = \tilde{\alpha}(\tilde{C} - \tilde{S}) \tag{M.3b}$$

$$\tilde{C}(X, T = 0) = 1, \quad X_w \le X \le X_b$$
$$\tilde{S}(X, T = 0) = 1, \quad X_w \le X \le X_b$$
$$\tilde{C}(X, T = 0) = S(X, T = 0) = 0, \quad X > X_b \tag{M.4a}$$

$$\tilde{C}(X = X_b, T) + \frac{\partial \tilde{C}(X = X_b, T)}{\partial X} = 0 \tag{M.4b}$$

$$\frac{\partial \tilde{C}(X = X_w, T)}{\partial X} = 0 \tag{M.4c}$$

where nondimensional variables have been defined as follows:

$$\tilde{C} = \frac{C}{C_0}, \quad \tilde{S} = \frac{S}{k_d C_0}, \quad X = \frac{r}{\alpha_r}, \quad T = \frac{Q_w t}{2\pi b\theta\alpha_r^2}, \quad \beta = \frac{\rho_b k_d}{\theta},$$

$$\tilde{\alpha} = \frac{2\pi b\theta\alpha_r^2\alpha}{Q_w}, \quad X_w = \frac{r_w}{\alpha_r}, \quad X_b = \frac{r_b}{\alpha_r}$$

These dimensionless equations are identical to those presented in Goltz and Oxley (1991). Following Goltz and Oxley (1991), we combine and transform Equations (M.3) using initial condition (M.4a) to obtain the following ordinary differential equation in the Laplace domain:

$$\frac{d^2\overline{C}}{dX^2} + \frac{d\overline{C}}{dX} - X\overline{C}\gamma = -X\frac{\gamma}{s} \tag{M.5a}$$

where $\gamma = s \left[1 + \frac{\beta\tilde{a}}{s+\tilde{a}} \right]$, s is the Laplace transform variable, and $\overline{C}(X,s)$ is the Laplace transform of the dimensionless concentration $\tilde{C}(X,T)$. Incidentally, for the linear equilibrium model, $\gamma = s$.

Transforming boundary conditions (Equations (M.4b) and (M.4c)) to the Laplace domain results in

$$\overline{C}(X = X_b, s) + \frac{d\overline{C}(X = X_b, s)}{dX} = 0 \qquad \text{(M.5b)}$$

$$\frac{d\overline{C}(X = X_w, s)}{dX} = 0 \qquad \text{(M.5c)}$$

Valocchi (1986) and Chen (1987) show how ordinary differential equation (M.5a) can be solved in the Laplace domain for the specified boundary conditions to obtain the following expression for concentration:

$$\overline{C}(X,s) = \frac{1}{s} \left\{ 1 - \exp[0.5(X^b - X)] \right.$$
$$\left. \bullet \left[\frac{Bi(y)G[Ai(y_w)] - Ai(y)G[Bi(y_w)]}{G[Ai(y_w)]H[Bi(y^b)] - G[Bi(y_w)]H[Ai(y^b)]} \right] \right\}$$

where

$$G[Ai(y)] = -0.5Ai(y) + \gamma^{1/3}\frac{d}{dy}Ai(y)$$

$$G[Bi(y)] = -0.5Bi(y) + \gamma^{1/3}\frac{d}{dy}Bi(y)$$

$$H[Ai(y)] = 0.5Ai(y) + \gamma^{1/3}\frac{d}{dy}Ai(y)$$

$$H[Bi(y)] = 0.5Bi(y) + \gamma^{1/3}\frac{d}{dy}Bi(y)$$

$$y = \gamma^{1/3}z, \quad z = X + (4\gamma)^{-1}$$
$$y_w = \gamma^{1/3}z_w, \quad z_w = X_w + (4\gamma)^{-1}$$
$$y_b = \gamma^{1/3}z_b, \quad z_b = X_b + (4\gamma)^{-1} \qquad \text{(M.6)}$$

and $Ai(y)$ and $Bi(y)$ are Airy functions.

References

Chen, C.S., Comment on "Effect of radial flow on deviations from local equilibrium during sorbing solute transport through homogeneous soils," by A.J. Valocchi, *Water Resources Research*, 23(11): 2157, 1987.

Chen, C.S. and G.D. Woodside, Analytical solution for aquifer decontamination by pumping, *Water Resources Research*, 24(8): 1329–1338, 1988.

Goltz, M.N. and M.E. Oxley, Analytical modeling of aquifer decontamination by pumping when transport is affected by rate-limited sorption, *Water Resources Research*, 27(4): 547–556, 1991.

Valocchi, A.J., Effect of radial flow on deviations from local equilibrium during sorbing solute transport through homogeneous soils, *Water Resources Research*, 22(12): 1693–1701, 1986.

N

AnaModelTool Governing Equations, Initial and Boundary Conditions, and Source Code

N.1 Model 101

Governing equations:

$$\frac{\partial C}{\partial t} + \frac{\rho_b}{\theta}\frac{\partial S}{\partial t} = D_x\frac{\partial^2 C}{\partial x^2} - v\frac{\partial C}{\partial x} - \lambda C \tag{N.101a}$$

$$\frac{\partial S}{\partial t} = \alpha(k_d C - S) \tag{N.101b}$$

Initial conditions:

$$C = S = 0,\ t = 0,\ 0 \le x \le L \tag{N.101c}$$

Boundary conditions:

$$C = C_0 H(t_s - t),\ x = 0,\ t > 0 \tag{N.101d}$$

$$C = 0,\ x = L,\ t > 0 \tag{N.101e}$$

Solution:

$$\overline{C}(x,s) = \frac{1}{s}C_0[1 - \exp(-t_s s)]\frac{\exp(r_2 L + r_1 x) - \exp(r_1 L + r_2 x)}{\exp(r_2 L) - \exp(r_1 L)} \tag{N.101f}$$

$$r_{1,2} = \frac{1}{2D_x}[v \pm (v^2 + 4D_x\Theta)^{1/2}] \tag{N.101g}$$

$$\Theta = s + \lambda + \frac{\rho_b \alpha k_d s}{\theta(s + \alpha)} \tag{N.101h}$$

Evaluation m-file:

```
function c=model_101(s,pv,x,sflag)

theta=pv(1);
rhob=pv(2);
```

Analytical Modeling of Solute Transport in Groundwater: Using Models to Understand the Effect of Natural Processes on Contaminant Fate and Transport, First Edition. Mark Goltz and Junqi Huang.
© 2017 John Wiley & Sons, Inc. Published 2017 by John Wiley & Sons, Inc.
Companion Website: www.wiley.com/go/Goltz/solute_transport_in_groundwater

```
D=pv(3);
v=pv(4);
lambda=pv(5);
alpha=pv(6);
kd=pv(7);
C0=pv(8);
L=pv(9);
ts=pv(10);

S=s*ones(size(x));
X=ones(size(s))*x;
if isinf(alpha)
    THETA=(S+lambda)+rhob/theta*kd*S;
else
    THETA=(S+lambda)+rhob/theta*alpha*kd*S./(S+alpha);
end

q=sqrt(v^2+4*D*THETA)/2/D;
e=v/2.0/D*X;
%slug=1-exp(-S*ts);
%slug is treated by superposition solutions in solver
slug=1;

f=(exp(-q.*(2*L-X)+e)-exp(-q.*X+e)) ./...
    (exp(-q*2*L)-1.0);
f(f==0)=1e-20;
c=C0 * slug ./ S .* f;

end
```

N.2   Model 102

Governing equations:

$$\frac{\partial C}{\partial t} + \frac{\rho_b}{\theta}\frac{\partial S}{\partial t} = D_x\frac{\partial^2 C}{\partial x^2} - v\frac{\partial C}{\partial x} - \lambda C \tag{N.102a}$$

$$\frac{\partial S}{\partial t} = \alpha(k_d C - S) \tag{N.102b}$$

Initial conditions:

$$C = S = 0, \; t = 0, \; 0 \le x \le L \tag{N.102c}$$

Boundary conditions:

$$C = C_0 H(t_s - t), \quad x = 0, \ t > 0 \tag{N.102d}$$

$$\partial C / \partial x = 0, \quad x = L, \ t > 0 \tag{N.102e}$$

Solution:

$$\overline{C}(x, s) = \frac{1}{s} C_0 [1 - \exp(-t_s s)]$$

$$\frac{r_2 \exp(r_2 L + r_1 x) - r_1 \exp(r_1 L + r_2 x)}{r_2 \exp(r_2 L) - r_1 \exp(r_1 L)} \tag{N.102f}$$

$$r_{1,2} = \frac{1}{2D_x} [v \pm (v^2 + 4D_x \Theta)^{1/2}] \tag{N.102g}$$

$$\Theta = s + \lambda + \frac{\rho_b \alpha k_d s}{\theta (s + \alpha)} \tag{N.102h}$$

Evaluation m-file:

```
function c=model_102(s,pv,x,sflag)

theta=pv(1);
rhob=pv(2);
D=pv(3);
v=pv(4);
lambda=pv(5);
alpha=pv(6);
kd=pv(7);
C0=pv(8);
L=pv(9);
ts=pv(10);

S=s*ones(size(x));
X=ones(size(s))*x;
if isinf(alpha)
    THETA=(S+lambda)+rhob/theta*kd*S;
else
    THETA=(S+lambda)+rhob/theta*alpha*kd*S./(S+alpha);
end
q=sqrt(v^2+4*D*THETA)/2/D;
e=v/2/D;
r1=e+q;
r2=e-q;

u=r2 .* exp(-2*q*L+r1.*X) - r1 .* exp(r2.*X);
w=r2 .* exp(-2*q*L) - r1;
```

```
%slug=1-exp(-s*ts);
%slug is treated by superposition solutions in solver
c=C0./S .* u ./ w;

end
```

N.3 Model 103

Governing equations:

$$\frac{\partial C}{\partial t} + \frac{\rho_b}{\theta}\frac{\partial S}{\partial t} = D_x\frac{\partial^2 C}{\partial x^2} - v\frac{\partial C}{\partial x} - \lambda C \tag{N.103a}$$

$$\frac{\partial S}{\partial t} = \alpha(k_d C - S) \tag{N.103b}$$

Initial conditions:

$$C = S = 0, \ t = 0, \ 0 \le x < \infty \tag{N.103c}$$

Boundary conditions:

$$C = C_0 H(t_s - t), \ x = 0, \ t > 0 \tag{N.103d}$$

$$C = 0, \ x \to \infty, \ t > 0 \tag{N.103e}$$

Solution:

$$\overline{C}(x,s) = \frac{1}{s}C_0[1 - \exp(-t_s s)]\exp(rx) \tag{N.103f}$$

$$r = \frac{1}{2D_x}[v - (v^2 + 4D_x\Theta)^{1/2}] \tag{N.103g}$$

$$\Theta = s + \lambda + \frac{\rho_b \alpha k_d s}{\theta(s + \alpha)} \tag{N.103h}$$

Evaluation m-file:

```
function c=model_103(s,pv,x,sflag)

theta=pv(1);
rhob=pv(2);
D=pv(3);
v=pv(4);
lambda=pv(5);
alpha=pv(6);
kd=pv(7);
C0=pv(8);
ts=pv(9);
```

```
S=s*ones(size(x));
X=ones(size(s))*x;
if isinf(alpha)
    THETA=(S+lambda)+rhob/theta*kd*S;
else
    THETA=(S+lambda)+rhob/theta*alpha*kd*S./(S+alpha);
end

q=sqrt(v^2+4*D*THETA);
r2=-2*THETA./(v+q);
%slug=1-exp(-s*ts);
%slug is treated by superposition solutions in solver

c=C0./S .* exp(r2.*X);

end
```

N.4   Model 104

Governing equations:

$$\frac{\partial C}{\partial t} + \frac{\rho_b}{\theta}\frac{\partial S}{\partial t} = D_x\frac{\partial^2 C}{\partial x^2} - v\frac{\partial C}{\partial x} - \lambda C \tag{N.104a}$$

$$\frac{\partial S}{\partial t} = \alpha(k_d C - S) \tag{N.104b}$$

Initial conditions:

$$C = M\delta(x), \ S = 0, \ t = 0, -\infty \leq x < \infty \tag{N.104c}$$

Boundary conditions:

$$C = 0, \ x \to \pm\infty, \ t > 0 \tag{N.104d}$$

Solution:

$$\overline{C}(x,s) = \frac{M}{2D_x\Delta}\exp(\sigma x - \Delta|x|) \tag{N.104e}$$

$$\sigma = \frac{v}{2D_x} \tag{N.104f}$$

$$\Delta = \frac{1}{2D_x}(v^2 + 4D_x\Theta)^{1/2} \tag{N.104g}$$

$$\Theta = s + \lambda + \frac{\rho_b\alpha k_d s}{\theta(s + \alpha)} \tag{N.104h}$$

Evaluation m-file:

```
function c=model_104(s,pv,x,sflag)

theta=pv(1);
rhob=pv(2);
D=pv(3);
v=pv(4);
lambda=pv(5);
alpha=pv(6);
kd=pv(7);
M0=pv(8);

S=s*ones(size(x));
X=ones(size(s))*x;
if isinf(alpha)
    THETA=(S+lambda)+rhob/theta*kd*S;
else
    THETA=(S+lambda)+rhob/theta*alpha*kd*S./(S+alpha);
end

q=sqrt(v^2+4*D*THETA)/2/D;
w=M0/2/D ./ q;
r=v/2/D;

c = w .* exp(r.*X-q.*abs(X));

end
```

N.5   Model 104M

Governing equations:

$$\frac{\partial C}{\partial t} + \frac{\rho_b}{\theta}\frac{\partial S}{\partial t} = D_x\frac{\partial^2 C}{\partial x^2} - v\frac{\partial C}{\partial x} - \lambda C \tag{N.104Ma}$$

$$\frac{\partial S}{\partial t} = \alpha(k_d C - S) \tag{N.104Mb}$$

Initial conditions:

$$C = M\delta(x), \; S = 0, \; t = 0, -\infty \leq x < \infty \tag{N.104Mc}$$

Boundary conditions:

$$\partial C/\partial x = 0, \quad x \to \pm\infty, \quad t > 0 \tag{N.104Md}$$

Solution:

$$\overline{C}(x, s) = \frac{M}{v - D_x r} e^{rx} \tag{N.104Me}$$

$$r = \frac{1}{2D_x}(v - \sqrt{v^2 + 4D_x\Theta}) \tag{N.104Mf}$$

$$\Theta = s + \lambda + \frac{\rho_b \alpha k_d s}{\theta(s + \alpha)} \tag{N.104Mg}$$

The zero-order temporal moment:

$$m_{0,t} = \frac{M}{v - D_x r_0} e^{r_0 x} \tag{N.104Mh}$$

The first-order temporal moment:

$$m_{1,t} = -M \left[\frac{D_x}{(v - D_x r_0)^2} + \frac{x}{(v - D_x r_0)} \right] r_0' e^{r_0 x} \tag{N.104Mi}$$

$$r_0 = \frac{1}{2D_x}(v - \sqrt{v^2 + 4D_x\lambda}) \tag{N.104Mj}$$

$$r_0' = -\frac{1}{\sqrt{v^2 + 4D_x\lambda}}(1 + \rho_b k_d/\theta) \tag{N.104Mk}$$

The zero-order spatial moment:

$$\overline{m}_{0,x} = -\frac{M}{(v - D_x r)r} = \frac{M}{\Theta} \tag{N.104Ml}$$

The first-order spatial moment:

$$\overline{m}_{1,x} = \frac{M}{(v - D_x r)r^2} = -\frac{M}{\Theta r} \tag{N.104Mm}$$

Evaluation m-file:

```
function val=model_104M(s,pv,j)

theta=pv(1);
rhob=pv(2);
D=pv(3);
v=pv(4);
lambda=pv(5);
```

```
alpha=pv(6);
kd=pv(7);
M=pv(8);
if isinf(alpha)
    THETA=(s+lambda)+rhob/theta*kd*s;
else
    THETA=(s+lambda)+rhob/theta*alpha*kd*s./(s+alpha);
end
if j==0
    val = M./THETA;
elseif j==1
    val = v*M*THETA.^(-2);
elseif j==2
    val=2*v^2*M * THETA.^(-3) + 2*D*M * THETA.^(-2);
end

end
```

N.6   Model 105

Governing equations:

$$\frac{\partial C}{\partial t} + \frac{\rho_b}{\theta}\frac{\partial S}{\partial t} = D_x\frac{\partial^2 C}{\partial x^2} - v\frac{\partial C}{\partial x} - \lambda C \tag{N.105a}$$

$$\frac{\partial S}{\partial t} = \alpha(k_d C - S) \tag{N.105b}$$

Initial conditions:

$$C = S = 0, \ t = 0, \ 0 \le x \le L \tag{N.105c}$$

Boundary conditions:

$$-D_x\partial C/\partial x + vC = vC_0 H(t_s - t), \ x = 0, \ t > 0 \tag{N.105d}$$

$$C = 0, \ x = L, \ t > 0 \tag{N.105e}$$

Solution:

$$\overline{C}(x, s) = \frac{1}{s}C_0[1 - \exp(-t_s s)]\frac{\exp(r_1 x - d) - \exp(r_2 x)}{(1 - hr_1)\exp(-d) - 1 + hr_2} \tag{N.105f}$$

$$r_{1,2} = \sigma \pm \Delta \tag{N.105g}$$

$$\sigma = \frac{v}{2D_x} \tag{N.105h}$$

$$\Delta = \frac{1}{2D_x}(v^2 + 4D_x\Theta)^{1/2} \tag{N.105i}$$

$$h = D_x/v \tag{N.105j}$$

$$d = 2\Delta L \tag{N.105k}$$

$$\Theta = s + \lambda + \frac{\rho_b \alpha k_d s}{\theta(s + \alpha)} \tag{N.105l}$$

Evaluation m-file:

```
function c=model_105(s,pv,x,sflag)

theta=pv(1);
rhob=pv(2);
D=pv(3);
v=pv(4);
lambda=pv(5);
alpha=pv(6);
kd=pv(7);
L=pv(8);
C0=pv(9);
ts=pv(10);

S=s*ones(size(x));
X=ones(size(s))*x;
if isinf(alpha)
    THETA=(S+lambda)+rhob/theta*kd*S;
else
    THETA=(S+lambda)+rhob/theta*alpha*kd*S./(S+alpha);
end

q=sqrt(v^2+4*D*THETA);
r1=(v+q)/2/D;
r2=(v-q)/2/D;
d=q/D*L;
h=D/v;

%slug=1-exp(-s*ts);
%slug is treated by superposition solutions in solver

f=(exp(r1.*X-d)-exp(r2.*X)) ./ ((1-h*r1)....
    *exp(-d)-1+h*r2);
f(f==0)=1e-10;
c=C0 ./ S .* f;

end
```

N.7   Model 106

Governing equations:

$$\frac{\partial C}{\partial t} + \frac{\rho_b}{\theta}\frac{\partial S}{\partial t} = D_x\frac{\partial^2 C}{\partial x^2} - v\frac{\partial C}{\partial x} - \lambda C \tag{N.106a}$$

$$\frac{\partial S}{\partial t} = \alpha(k_d C - S) \tag{N.106b}$$

Initial conditions:

$$C = S = 0, \ t = 0, \ 0 \le x \le L \tag{N.106c}$$

Boundary conditions:

$$-D_x \partial C/\partial x + vC = vC_0 H(t_s - t), \ x = 0, \ t > 0 \tag{N.106d}$$

$$\partial C/\partial x = 0, \ x = L, \ t > 0 \tag{N.106e}$$

Solution:

$$\overline{C}(x,s) = \frac{1}{s}C_0[1 - \exp(-t_s s)]\frac{r_2\exp(r_1 x - d) - r_1\exp(r_2 x)}{r_2(1 - hr_1)\exp(-d) - (1 - hr_2)r_1} \tag{N.106f}$$

$$r_{1,2} = \sigma \pm \Delta \tag{N.106g}$$

$$\sigma = \frac{v}{2D_x} \tag{N.106h}$$

$$\Delta = \frac{1}{2D_x}(v^2 + 4D_x\Theta)^{1/2} \tag{N.106i}$$

$$h = D_x/v \tag{N.106j}$$

$$d = 2\Delta L \tag{N.106k}$$

$$\Theta = s + \lambda + \frac{\rho_b \alpha k_d s}{\theta(s + \alpha)} \tag{N.106l}$$

Evaluation m-file:

```
function c=model_106(s,pv,x,sflag)

theta=pv(1);
rhob=pv(2);
D=pv(3);
v=pv(4);
lambda=pv(5);
alpha=pv(6);
kd=pv(7);
```

```
L=pv(8);
C0=pv(9);
ts=pv(10);

S=s*ones(size(x));
X=ones(size(s))*x;
if isinf(alpha)
    THETA=(S+lambda)+rhob/theta*kd*S;
else
    THETA=(S+lambda)+rhob/theta*alpha*kd*S./(S+alpha);
end

q=sqrt(v^2+4*D*THETA);
r1=(v+q)/2/D;
r2=(v-q)/2/D;
d=q/D*L;
h=D/v;

%slug=1-exp(-s*ts);
%slug is treated by superposition solutions in solver

c=C0 ./ S .* (r2.*exp(r1.*X-d)-r1.*exp(r2.*X)) ./...
(r2.*(1-h*r1).*exp(-d)-(1-h*r2).*r1);

end
```

N.8   Model 107

Governing equations:

$$\frac{\partial C}{\partial t} + \frac{\rho_b}{\theta}\frac{\partial S}{\partial t} = D_x\frac{\partial^2 C}{\partial x^2} - v\frac{\partial C}{\partial x} - \lambda C \tag{N.107a}$$

$$\frac{\partial S}{\partial t} = \alpha(k_d C - S) \tag{N.107b}$$

Initial conditions:

$$C = S = 0,\ t = 0,\ 0 \leq x < \infty \tag{N.107c}$$

Boundary conditions:

$$- D_x\partial C/\partial x + vC = vC_0 H(t_s - t),\ x = 0,\ t > 0 \tag{N.107d}$$

$$C = 0,\ x \to \infty,\ t > 0 \tag{N.107e}$$

Solution:

$$\overline{C}(x,s) = \frac{1}{s}C_0[1 - \exp(-t_s s)]\frac{\exp(rx)}{1 - hr} \qquad \text{(N.107f)}$$

$$r = \frac{1}{2D_x}(v - \sqrt{v^2 + 4D_x\Theta}) \qquad \text{(N.107g)}$$

$$h = D_x/v \qquad \text{(N.107h)}$$

$$\Theta = s + \lambda + \frac{\rho_b \alpha k_d s}{\theta(s + \alpha)} \qquad \text{(N.107i)}$$

Evaluation m-file:

```
function c=model_107(s,pv,x,sflag)

theta=pv(1);
rhob=pv(2);
D=pv(3);
v=pv(4);
lambda=pv(5);
alpha=pv(6);
kd=pv(7);
C0=pv(8);
ts=pv(9);

S=s*ones(size(x));
X=ones(size(s))*x;
if isinf(alpha)
    THETA=(S+lambda)+rhob/theta*kd*S;
else
    THETA=(S+lambda)+rhob/theta*alpha*kd*S./(S+alpha);
end
q=sqrt(v^2+4*D*THETA);
r2=(v-q)/2.0/D;
h=D/v;

%slug=1-exp(-s*ts);
%slug is treated by superposition solutions in solver

c=C0 ./ S .* exp(r2.*X) ./(1-h*r2);

end
```

N.9 Model 108

Governing equations:

$$\frac{\partial C}{\partial t} + \frac{\rho_b}{\theta}\frac{\partial S}{\partial t} = D_x\frac{\partial^2 C}{\partial x^2} - v\frac{\partial C}{\partial x} - \lambda C + \frac{2D_m}{b}\frac{\partial C_m}{\partial z}\bigg|_{z=0} \tag{N.108a}$$

$$\frac{\partial S}{\partial t} = \alpha(k_d C - S) \tag{N.108b}$$

$$\frac{\partial C_m}{\partial t} + \frac{\rho_m}{\theta_m}\frac{\partial S_m}{\partial t} = D_m\frac{\partial^2 C}{\partial z^2} - \lambda C_m \tag{N.108c}$$

$$\frac{\partial S_m}{\partial t} = \alpha_m(k_m C_m - S_m) \tag{N.108d}$$

Initial conditions:

$$C = S = 0,\ t = 0,\ 0 \le x < \infty \tag{N.108e}$$

$$C_m = S_m = 0,\ t = 0,\ 0 \le z < \infty \tag{N.108f}$$

Boundary conditions:

$$C = C_0 H(t_s - t),\ x = 0,\ t > 0 \tag{N.108g}$$

$$C_m = C,\ z = 0,\ 0 \le x < \infty,\ t > 0 \tag{N.108h}$$

$$C = 0,\ x \to \infty,\ t > 0 \tag{N.108i}$$

$$C_m = 0,\ z \to \infty,\ t > 0 \tag{N.108j}$$

Solution:

$$\overline{C}(x,s) = \frac{1}{s}C_0[1 - \exp(-t_s s)]\exp(rx) \tag{N.108k}$$

$$r = \frac{1}{2D_x}[v - \sqrt{v^2 + 4D_x(\Theta + \Gamma)}] \tag{N.108l}$$

$$\Theta = s + \lambda + \frac{\rho_b \alpha k_d s}{\theta(s + \alpha)} \tag{N.107m}$$

$$\Gamma = \frac{2}{b}\sqrt{D_m \Theta_m} \tag{N.108n}$$

$$\Theta_m = s + \lambda + \frac{\rho_m \alpha_m k_m s}{\theta_m(s + \alpha_m)} \tag{N.108o}$$

Evaluation m-file:

```
function c=model_108(s,pv,x,sflag)

theta=pv(1);
```

```
rhob=pv(2);
D=pv(3);
v=pv(4);
lambda=pv(5);
alpha=pv(6);
kd=pv(7);
C0=pv(8);
thetam=pv(9);
rhom=pv(10);
Dm=pv(11);
alpham=pv(12);
km=pv(13);
b=pv(14);
ts=pv(15);

S=s*ones(size(x));
X=ones(size(s))*x;
if isinf(alpha)
     THETA=(S+lambda)+rhob/theta*kd*S;
else
     THETA=(S+lambda)+rhob/theta*alpha*kd*S./...
     (S+alpha);
end
if isinf(alpham)
THETAM=(S+alpham)+rhom/thetam*km*S;
else
     THETAM=(S+alpham)+rhom/thetam*alpham*km*S./...
     (S+alpham);
end
GAMMA=2*sqrt(Dm*THETAM)/b;

delt=THETA +GAMMA;
q=sqrt(v^2+4*D*delt);
r2=-2*delt./(v+q);

%slug=1-exp(-s*ts);
%slug is treated by superposition solutions in...
     solver

c=C0./S .* exp(r2.*X);

end
```

N.10   Model 109

Governing equations:

$$\frac{\partial C}{\partial t} + \frac{\rho_b}{\theta}\frac{\partial S}{\partial t} = D_x\frac{\partial^2 C}{\partial x^2} - v\frac{\partial C}{\partial x} - \lambda C + \frac{2D_m}{b}\frac{\partial C_m}{\partial z}\bigg|_{z=0} \tag{N.109a}$$

$$\frac{\partial S}{\partial t} = \alpha(k_d C - S) \tag{N.109b}$$

$$\frac{\partial C_m}{\partial t} + \frac{\rho_m}{\theta_m}\frac{\partial S_m}{\partial t} = D_m\frac{\partial^2 C}{\partial z^2} - \lambda C_m \tag{N.109c}$$

$$\frac{\partial S_m}{\partial t} = \alpha_m(k_m C_m - S_m) \tag{N.109d}$$

Initial conditions:

$$C = S = 0, \ t = 0, \ 0 \le x < \infty \tag{N.109e}$$

$$C_m = S_m = 0, \ t = 0, \ 0 \le z < \infty \tag{N.109f}$$

Boundary conditions:

$$-D_x \partial C/\partial x + vC = vC_0 H(t_s - t), \ x = 0, \ t > 0 \tag{N.109g}$$

$$C_m = C, \ z = 0, \ 0 \le x < \infty, \ t > 0 \tag{N.109h}$$

$$C = 0, \ x \to \infty, \ t > 0 \tag{N.109i}$$

$$C_m = 0, \ z \to \infty, \ t > 0 \tag{N.109j}$$

Solution:

$$\overline{C}(x,s) = \frac{1}{s}C_0[1 - \exp(-t_s s)]\frac{\exp(rx)}{1 - hr} \tag{N.109k}$$

$$r = \frac{1}{2D_x}[v - \sqrt{v^2 + 4D_x(\Theta + \Gamma)}] \tag{N.109l}$$

$$\Theta = s + \lambda + \frac{\rho_b \alpha k_d s}{\theta(s + \alpha)} \tag{N.109m}$$

$$\Gamma = \frac{2}{b}\sqrt{D_m \Theta_m} \tag{N.109n}$$

$$\Theta_m = s + \lambda + \frac{\rho_m \alpha_m k_m s}{\theta_m(s + \alpha_m)} \tag{N.109o}$$

$$h = D_x/v \tag{N.109p}$$

Evaluation m-file:

```
function c=model_109(s,pv,x,sflag)

theta=pv(1);
rhob=pv(2);
D=pv(3);
v=pv(4);
lambda=pv(5);
alpha=pv(6);
kd=pv(7);
C0=pv(8);
thetam=pv(9);
rhom=pv(10);
Dm=pv(11);
alpham=pv(12);
km=pv(13);
b=pv(14);
ts=pv(15);

S=s*ones(size(x));
X=ones(size(s))*x;
if isinf(alpha)
    THETA=(S+lambda)+rhob/theta*kd*S;
else
    THETA=(S+lambda)+rhob/theta*alpha*kd*S./(S+alpha);
end
if isinf(alpham)
    THETAM=(S+alpham)+rhom/thetam*km*S;
else
    THETAM=(S+alpham)+rhom/thetam*alpham*km*S./...
    (S+alpham);
end
GAMMA=2*sqrt(Dm*THETAM)/b;

delt=THETA +GAMMA;
q=sqrt(v^2+4*D*delt);
r2=(v-q)/2.0/D;
h=D/v;

%slug=1-exp(-s*ts);
```

```
%slug is treated by superposition solutions in solver

c=C0 ./ S .* exp(r2.*X) ./(1-h*r2);

end
```

N.11 Model 201

Governing equations:

$$\frac{\partial C}{\partial t} + \frac{\rho_b}{\theta}\frac{\partial S}{\partial t} = D_x\frac{\partial^2 C}{\partial x^2} - v\frac{\partial C}{\partial x} + D_y\frac{\partial^2 C}{\partial y^2} - \lambda C \tag{N.201a}$$

$$\frac{\partial S}{\partial t} = \alpha(k_d C - S) \tag{N.201b}$$

Initial conditions:

$$C = S = 0, \ t = 0, \ 0 \le x \le a, \ 0 \le y \le b \tag{N.201c}$$

Boundary conditions:

$$C = \begin{cases} C_0 H(t_s - t), & y_b \le y \le y_t \\ 0, & y_b > y > y_t \end{cases}, \ x = 0, \ t > 0 \tag{N.201d}$$

$$C = 0, \ x = a, \ 0 \le y \le b, \ t > 0 \tag{N.201e}$$

$$C = 0, \ y = 0, b, \ 0 \le x \le a, \ t > 0 \tag{N.201f}$$

Solution:

$$\overline{C}(x, s) = \frac{2}{b}C_0[1 - \exp(-t_s s)]$$

$$\times \sum_{n=1}^{\infty} \gamma\frac{\exp(r_2 a + r_1 x) - \exp(r_1 a + r_2 x)}{\exp(r_2 a) - \exp(r_1 a)}\sin\left(\frac{n\pi y}{b}\right) \tag{N.201g}$$

$$\gamma = \frac{b}{sn\pi}[\cos(n\pi y_b/b) - \cos(n\pi y_t/b) \tag{N.201h}$$

$$r_{1,2} = \frac{1}{2D_x}\{v \pm [v^2 + 4D_x[D_y(n\pi/b)^2 + \Theta]^{1/2}\} \tag{N.201i}$$

$$\Theta = s + \lambda + \frac{\rho_b \alpha k_d s}{\theta(s + \alpha)} \tag{N.201j}$$

Evaluation m-file:

```
function c=model_201(s,pv,x,y)

theta=pv(1);
rhob=pv(2);
Dx=pv(3);
vx=pv(4);
Dy=pv(5);
lambda=pv(6);
alpha=pv(7);
kd=pv(8);
a=pv(9);
b=pv(10);
C0=pv(11);
ts=pv(12);
yb=pv(13);
yt=pv(14);
N=pv(15);
S=s*ones(size(x));
X=ones(size(s))*x;
if isinf(alpha)
THETA=(S+lambda)+rhob/theta*kd*S;
else
THETA=(S+lambda)+rhob/theta*alpha*kd*S./(S+alpha);
end
%slug=1-exp(-s*ts);
%slug is treated by superposition solutions in solver
NN=(1:N);
N2=(NN*pi/b).^2;
SS=sin(NN*pi*y/b);

gamma=C0*b/pi * (cos(NN*pi*yb/b)-cos(NN*pi*yt/b)) ./...
    NN;
tt=0;
for k=1:N
q=sqrt(vx^2+4*Dx*(Dy*N2(k)+THETA));
r1=(vx+q)/2.0/Dx;
r2=(vx-q)/2.0/Dx;
if isinf(a)
ee=exp(r2.*X);
else
ee=(exp((r2-r1)*a+r1.*X)-exp(r2.*X))./...
    (exp((r2-r1)*a)-1.);
```

```
end
tt=tt+SS(k)*gamma(k)  .* ee;
end
f  =  2.0/b * tt;
f=f+(f==0)*1e-15;
c=  f./S;
end
```

N.12   Model 202

Governing equations:

$$\frac{\partial C}{\partial t} + \frac{\rho_b}{\theta}\frac{\partial S}{\partial t} = D_x\frac{\partial^2 C}{\partial x^2} - v\frac{\partial C}{\partial x} + D_y\frac{\partial^2 C}{\partial y^2} - \lambda C \qquad \text{(N.202a)}$$

$$\frac{\partial S}{\partial t} = \alpha(k_d C - S) \qquad \text{(N.202b)}$$

Initial conditions:

$$C = S = 0,\ t = 0,\ 0 \le x \le a,\ 0 \le y \le b \qquad \text{(N.202c)}$$

Boundary conditions:

$$C = \begin{cases} C_0 H(t_s - t), & y_b \le y \le y_t \\ 0, & y_b > y > y_t \end{cases},\ x = 0,\ t > 0 \qquad \text{(N.202d)}$$

$$C = 0,\ x = a,\ 0 \le y \le b,\ t > 0 \qquad \text{(N.202e)}$$

$$\partial C/\partial y = 0,\ y = 0, b,\ 0 \le x \le a,\ t > 0 \qquad \text{(N.202f)}$$

Solution:

$$\overline{C}(x,s) = \frac{1}{b}C_0[1 - \exp(-t_s s)]\left[\hat{\overline{C}}(0) + 2\sum_{n=1}^{\infty}\hat{\overline{C}}(n)\cos(n\pi y/b)\right] \qquad \text{(N.202g)}$$

$$\hat{\overline{C}}(n) = \gamma\frac{\exp(r_2 a + r_1 x) - \exp(r_1 a + r_2 x)}{\exp(r_2 a) - \exp(r_1 a)} \qquad \text{(N.202h)}$$

$$\gamma = \frac{C_0 b}{s n \pi}[\sin(n\pi y_t/b) - \sin(n\pi y_b/b)] \qquad \text{(N.202i)}$$

$$r_{1,2} = \frac{1}{2D_x}\{v \pm [v^2 + 4D_x[D_y(n\pi/b)^2 + \Theta]^{1/2}\} \qquad \text{(N.202j)}$$

$$\Theta = s + \lambda + \frac{\rho_b \alpha k_d s}{\theta(s + \alpha)} \qquad \text{(N.202k)}$$

Evaluation m-file:

```
function c=model_202(s,pv,x,y)

theta=pv(1);
rhob=pv(2);
Dx=pv(3);
vx=pv(4);
Dy=pv(5);
lambda=pv(6);
alpha=pv(7);
kd=pv(8);
a=pv(9);
b=pv(10);
C0=pv(11);
ts=pv(12);
yb=pv(13);
yt=pv(14);
N=pv(15);

S=s*ones(size(x));
X=ones(size(s))*x;
if isinf(alpha)
    THETA=(S+lambda)+rhob/theta*kd*S;
else
    THETA=(S+lambda)+rhob/theta*alpha*kd*S./(S+alpha);
end

q=sqrt(vx^2+4*Dx*THETA);
r01=(vx+q)/2.0/Dx;
r02=(vx-q)/2.0/Dx;

if isinf(a)
    f0=exp(r02.*X);
else
    f0=(exp((r02-r01)*a+r01.*X)-exp(r02.*X))./...
    (exp((r02-r01)*a)-1.);
end
t0=C0*(yt-yb)/b * f0;

%slug=1-exp(-s*ts);
%slug is treated by superposition solutions in solver
```

```
NN=(1:N);
N2=(NN*pi/b).^2;
CC=cos(NN*pi*y/b);
gamma=C0*b/pi * (sin(NN*pi*yt/b)-sin(NN*pi*yb/b)) ./...
   NN;
tt=zeros(size(t0));

for k=1:N
    q=sqrt(vx^2+4*Dx*(Dy*N2(k)+THETA));
    r1=(vx+q)/2.0/Dx;
    r2=(vx-q)/2.0/Dx;
    if isinf(a)
        ee=exp(r2.*X);
    else
    ee=(exp((r2-r1)*a+r1.*X)-exp(r2.*X))./...
    (exp((r2-r1)*a)-1.);
    end
    tt=tt+CC(k)*gamma(k)  .* ee;
end
f = t0 + 2.0/b * tt;
f=f+(f==0)*1e-15;
c= f./S;

end
```

N.13   Model 203

Governing equations:

$$\frac{\partial C}{\partial t} + \frac{\rho_b}{\theta}\frac{\partial S}{\partial t} = D_x\frac{\partial^2 C}{\partial x^2} - v\frac{\partial C}{\partial x} + D_y\frac{\partial^2 C}{\partial y^2} - \lambda C \qquad \text{(N.203a)}$$

$$\frac{\partial S}{\partial t} = \alpha(k_d C - S) \qquad \text{(N.203b)}$$

Initial conditions:

$$C = S = 0, \ t = 0, \ 0 \le x \le a, \ 0 \le y \le b \qquad \text{(N.203c)}$$

Boundary conditions:

$$C = \begin{cases} C_0 H(t_s - t), & y_b \le y \le y_t \\ 0, & y_b > y > y_t \end{cases}, \ x = 0, \ t > 0 \qquad \text{(N.203d)}$$

$$\partial C/\partial x = 0, \ x = a, \ 0 \le y \le b, \ t > 0 \qquad \text{(N.203e)}$$

$$C = 0, \ y = 0, b, \ 0 \le x \le a, \ t > 0 \tag{N.203f}$$

Solution:

$$\overline{C}(x,s) = \frac{2C_0}{b}[1 - \exp(-t_s s)] \sum_{n=1}^{\infty} \hat{C}(n) \sin(n\pi y/b)] \tag{N.203g}$$

$$\hat{C}(n) = \gamma \frac{r_2 \exp(r_2 a + r_1 x) - r_1 \exp(r_1 a + r_2 x)}{r_2 \exp(r_2 a) - r_1 \exp(r_1 a)} \tag{N.203h}$$

$$\gamma = \frac{b}{sn\pi}[\cos(n\pi y_b/b) - \cos(n\pi y_t/b) \tag{N.203i}$$

$$r_{1,2} = \frac{1}{2D_x}\{v \pm [v^2 + 4D_x[D_y(n\pi/b)^2 + \Theta]^{1/2}\} \tag{N.203j}$$

$$\Theta = s + \lambda + \frac{\rho_b \alpha k_d s}{\theta(s + \alpha)} \tag{N.203k}$$

Evaluation m-file:

```
function c=model_203(s,pv,x,y)

theta=pv(1);
rhob=pv(2);
Dx=pv(3);
vx=pv(4);
Dy=pv(5);
lambda=pv(6);
alpha=pv(7);
kd=pv(8);
a=pv(9);
b=pv(10);
C0=pv(11);
ts=pv(12);
yb=pv(13);
yt=pv(14);
N=pv(15);

S=s*ones(size(x));
X=ones(size(s))*x;
if isinf(alpha)
    THETA=(S+lambda)+rhob/theta*kd*S;
else
    THETA=(S+lambda)+rhob/theta*alpha*kd*S./(S+alpha);
end
```

```
%slug=1-exp(-s*ts);
%slug is treated by superposition solutions in solver
NN=(1:N);
N2=(NN*pi/b).^2;
SS=sin(NN*pi*y/b);
gamma=C0*b/pi * (cos(NN*pi*yb/b)-cos(NN*pi*yt/b)) ./...
    NN;
tt=0;

for k=1:N
    q=sqrt(vx^2+4*Dx*(Dy*N2(k)+THETA));
    r1=(vx+q)/2.0/Dx;
    r2=(vx-q)/2.0/Dx;
    if isinf(a)
        ee=exp(r2.*X);
    else
        ee=(r2.*exp((r2-r1)*a+r1.*X)-r1.*...
            exp(r2.*X))./(r2.*exp((r2-r1)*a)-r1);
    end
    tt=tt+SS(k)*gamma(k) .* ee;
end

f = 2.0/b * tt;
f=f+(f==0)*1e-15;
c= f./S;

end
```

N.14   Model 204

Governing equations:

$$\frac{\partial C}{\partial t} + \frac{\rho_b}{\theta}\frac{\partial S}{\partial t} = D_x\frac{\partial^2 C}{\partial x^2} - v\frac{\partial C}{\partial x} + D_y\frac{\partial^2 C}{\partial y^2} - \lambda C \tag{N.204a}$$

$$\frac{\partial S}{\partial t} = \alpha(k_d C - S) \tag{N.204b}$$

Initial conditions:

$$C = S = 0,\ t = 0,\ 0 \le x \le a,\ 0 \le y \le b \tag{N.204c}$$

Boundary conditions:

$$C = \begin{cases} C_0 H(t_s - t), & y_b \le y \le y_t \\ 0, & y_b > y > y_t \end{cases},\ x = 0,\ t > 0 \tag{N.204d}$$

$$\partial C/\partial x = 0, \quad x = a, \; 0 \le y \le b, \; t > 0 \tag{N.204e}$$

$$\partial C/\partial y = 0, \quad y = 0, b, \; 0 \le x \le a, \; t > 0 \tag{N.204f}$$

Solution:

$$\overline{C}(x,s) = \frac{1}{b}[1 - \exp(-t_s s)]\left[\hat{\overline{C}}(0) + 2\sum_{n=1}^{\infty} \hat{\overline{C}}(n)\cos(n\pi y/b)\right] \tag{N.204g}$$

$$\hat{\overline{C}}(n) = \gamma \frac{r_2 \exp(r_2 a + r_1 x) - r_1 \exp(r_1 a + r_2 x)}{r_2 \exp(r_2 a) - r_1 \exp(r_1 a)} \tag{N.204h}$$

$$\gamma = \frac{C_0 b}{sn\pi}[\sin(n\pi y_t/b) - \sin(n\pi y_b/b)] \tag{N.204i}$$

$$r_{1,2} = \frac{1}{2D_x}\{v \pm [v^2 + 4D_x[D_y(n\pi/b)^2 + \Theta]^{1/2}\} \tag{N.204j}$$

$$\Theta = s + \lambda + \frac{\rho_b \alpha k_d s}{\theta(s + \alpha)} \tag{N.204k}$$

Evaluation m-file:

```
function c=model_204(s,pv,x,y)

theta=pv(1);
rhob=pv(2);
Dx=pv(3);
vx=pv(4);
Dy=pv(5);
lambda=pv(6);
alpha=pv(7);
kd=pv(8);
a=pv(9);
b=pv(10);
C0=pv(11);
ts=pv(12);
yb=pv(13);
yt=pv(14);
N=pv(15);

S=s*ones(size(x));
X=ones(size(s))*x;
if isinf(alpha)
```

```
    THETA=(S+lambda)+rhob/theta*kd*S;
else
    THETA=(S+lambda)+rhob/theta*alpha*kd*S./(S+alpha);
end

q=sqrt(vx^2+4*Dx*THETA);
r01=(vx+q)/2.0/Dx;
r02=(vx-q)/2.0/Dx;

if isinf(a)
    f0=exp(r02.*X);
else
    f0=(r02.*exp((r02-r01)*a+r01.*X)-r01.*...
        exp(r02.*X))./ (r02.*exp((r02-r01)*a)-r01);
end
t0=C0*(yt-yb)/b * f0;

%slug=1-exp(-s*ts);
%slug is treated by superposition solutions in solver
NN=(1:N);
N2=(NN*pi/b).^2;
CC=cos(NN*pi*y/b);
gamma=C0*b/pi*(sin(NN*pi*yt/b)-sin(NN*pi*yb/b))./...
    NN;
tt=zeros(size(t0));

for k=1:N
    q=sqrt(vx^2+4*Dx*(Dy*N2(k)+THETA));
    r1=(vx+q)/2.0/Dx;
    r2=(vx-q)/2.0/Dx;
    if isinf(a)
      ee=exp(r2.*X);
    else
      ee=(r2.*exp((r2-r1)*a+r1.*X)-r1.*exp(r2.*X)) ...
          ./ (r2.*exp((r2-r1)*a)-r1);
    end
    tt=tt+CC(k)*gamma(k) .* ee;
end
f = t0 + 2.0/b * tt;
f=f+(f==0)*1e-15;
c= f./S;
end
```

N.15   Model 205

Governing equations:

$$\frac{\partial C}{\partial t} + \frac{\rho_b}{\theta}\frac{\partial S}{\partial t} = D_x\frac{\partial^2 C}{\partial x^2} - v\frac{\partial C}{\partial x} + D_y\frac{\partial^2 C}{\partial y^2} - \lambda C \qquad \text{(N.205a)}$$

$$\frac{\partial S}{\partial t} = \alpha(k_d C - S) \qquad \text{(N.205b)}$$

Initial conditions:

$$C = S = 0, \ t = 0, \ 0 \le x \le a, \ 0 \le y \le b \qquad \text{(N.205c)}$$

Boundary conditions:

$$C = \begin{cases} C_0 H(t_s - t), & y_b \le y \le y_t \\ 0, & y_b > y > y_t \end{cases}, \ x = 0, \ t > 0 \qquad \text{(N.205d)}$$

$$C = 0, \ x \to \infty, \ 0 \le y \le b, \ t > 0 \qquad \text{(N.205e)}$$

$$C = 0, \ y = 0, b, \ 0 \le x < \infty, \ t > 0 \qquad \text{(N.205f)}$$

Solution:

$$\overline{C}(x,s) = \frac{2}{b}C_0[1 - \exp(-t_s s)]\sum_{n=1}^{\infty} \gamma \exp(rx) \sin\left(\frac{n\pi y}{b}\right) \qquad \text{(N.205g)}$$

$$\gamma = \frac{b}{sn\pi}[\cos(n\pi y_b/b) - \cos(n\pi y_t/b) \qquad \text{(N.205h)}$$

$$r = \frac{1}{2D_x}\{v - [v^2 + 4D_x[D_y(n\pi/b)^2 + \Theta]^{1/2}\} \qquad \text{(N.205i)}$$

$$\Theta = s + \lambda + \frac{\rho_b \alpha k_d s}{\theta(s + \alpha)} \qquad \text{(N.205j)}$$

Evaluation m-file:

```
function c=model_205(s,pv,x,y)

theta=pv(1);
rhob=pv(2);
Dx=pv(3);
vx=pv(4);
Dy=pv(5);
lambda=pv(6);
alpha=pv(7);
kd=pv(8);
b=pv(9);
C0=pv(10);
ts=pv(11);
```

```
yb=pv(12);
yt=pv(13);
N=pv(14);

S=s*ones(size(x));
X=ones(size(s))*x;
if isinf(alpha)
    THETA=(S+lambda)+rhob/theta*kd*S;
else
    THETA=(S+lambda)+rhob/theta*alpha*kd*S./(S+alpha);
end

%slug=1-exp(-s*ts);
%slug is treated by superposition solutions in solver

NN=(1:N);
N2=(NN*pi/b).^2;
SS=sin(NN*pi*y/b);
gamma=C0*b/pi*(cos(NN*pi*yb/b)-cos(NN*pi*yt/b))./...
    NN;
tt=0;

for k=1:N
    q=sqrt(vx^2+4*Dx*(Dy*N2(k)+THETA));
    r2=(vx-q)/2.0/Dx;
    ee=exp(r2.*X);
    tt=tt+SS(k)*gamma(k)  .* ee;
end
f = 2.0/b * tt;
f=f+(f==0)*1e-15;
c= f./S;

end
```

N.16   Model 206

Governing equations:

$$\frac{\partial C}{\partial t} + \frac{\rho_b}{\theta}\frac{\partial S}{\partial t} = D_x\frac{\partial^2 C}{\partial x^2} - v\frac{\partial C}{\partial x} + D_y\frac{\partial^2 C}{\partial y^2} - \lambda C \tag{N.206a}$$

$$\frac{\partial S}{\partial t} = \alpha(k_d C - S) \tag{N.206b}$$

Initial conditions:

$$C = S = 0, \ t = 0, \ 0 \le x \le a, \ 0 \le y \le b \qquad \text{(N.206c)}$$

Boundary conditions:

$$C = \begin{cases} C_0 H(t_s - t), & y_b \le y \le y_t \\ 0, & y_b > y > y_t \end{cases}, \ x = 0, \ t > 0 \qquad \text{(N.206d)}$$

$$C = 0, \ x \to \infty, \ 0 \le y \le b, \ t > 0 \qquad \text{(N.206e)}$$

$$\partial C / \partial y = 0, \ y = 0, b, \ 0 \le x < \infty, \ t > 0 \qquad \text{(N.206f)}$$

Solution:

$$\overline{C}(x,s) = \frac{1}{b} C_0 [1 - \exp(-t_s s)] \left[\hat{\overline{C}}(0) + 2 \sum_{n=1}^{\infty} \hat{\overline{C}}(n) \cos(n\pi y / b) \right] \quad \text{(N.206g)}$$

$$\hat{\overline{C}}(n) = \gamma \exp(rx) \qquad \text{(N.206h)}$$

$$\gamma = \frac{C_0 b}{s n \pi} [\sin(n\pi y_t / b) - \sin(n\pi y_b / b) \qquad \text{(N.206i)}$$

$$r = \frac{1}{2D_x} \{ v - [v^2 + 4D_x [D_y (n\pi / b)^2 + \Theta]^{1/2} \} \qquad \text{(N.206j)}$$

$$\Theta = s + \lambda + \frac{\rho_b \alpha k_d s}{\theta(s + \alpha)} \qquad \text{(N.206k)}$$

Evaluation m-file:

```
function c=model_206(s,pv,x,y)

theta=pv(1);
rhob=pv(2);
Dx=pv(3);
vx=pv(4);
Dy=pv(5);
lambda=pv(6);
alpha=pv(7);
kd=pv(8);
b=pv(9);
C0=pv(10);
ts=pv(11);
yb=pv(12);
yt=pv(13);
N=pv(14);
```

```
S=s*ones(size(x));
X=ones(size(s))*x;
if isinf(alpha)
    THETA=(S+lambda)+rhob/theta*kd*S;
else
    THETA=(S+lambda)+rhob/theta*alpha*kd*S./(S+alpha);
end

q=sqrt(vx^2+4*Dx*THETA);
r02=(vx-q)/2.0/Dx;

f0=exp(r02.*X);
t0=C0*(yt-yb)/b * f0;

%slug=1-exp(-s*ts);
%slug is treated by superposition solutions in solver

NN=(1:N);
N2=(NN*pi/b).^2;
CC=cos(NN*pi*y/b);
gamma=C0*b/pi*(sin(NN*pi*yt/b)-sin(NN*pi*yb/b))./...
    NN;
tt=zeros(size(t0));

for k=1:N
    q=sqrt(vx^2+4*Dx*(Dy*N2(k)+THETA));
    r2=(vx-q)/2.0/Dx;
    ee=exp(r2.*X);
    tt=tt+CC(k)*gamma(k)  .* ee;
end
f = t0 + 2.0/b * tt;
f(f==0)=1e-15;
c= f./S;

end
```

N.17 Model 207

Governing equations:

$$\frac{\partial C}{\partial t} + \frac{\rho_b}{\theta}\frac{\partial S}{\partial t} = D_x \frac{\partial^2 C}{\partial x^2} - v\frac{\partial C}{\partial x} + D_y \frac{\partial^2 C}{\partial y^2} - \lambda C \qquad (N.207a)$$

$$\frac{\partial S}{\partial t} = \alpha(k_d C - S) \tag{N.207b}$$

Initial conditions:

$$C = S = 0, \; t = 0, \; 0 \le x \le a, \; 0 \le y \le b \tag{N.207c}$$

Boundary conditions:

$$C = \begin{cases} C_0 H(t_s - t), & y_b \le y \le y_t \\ 0, & y_b > y > y_t \end{cases}, \; x = 0, \; t > 0 \tag{N.207d}$$

$$C = 0, \; x \to \infty, \; -\infty < y < \infty, \; t > 0 \tag{N.207e}$$

$$\partial C/\partial y = 0, \; y = \pm\infty, \; 0 \le x < \infty, \; t > 0 \tag{N.207f}$$

Solution:

$$\overline{C}(x,s) = \gamma x \exp\left(\frac{vx}{2D_x}\right) \int_{-b}^{b} \frac{K_1[\beta \sqrt{(x^2/D_x + (y - \xi)^2/D_y)}]}{\sqrt{(x^2/D_x + (y - \xi)^2/D_y)}} d\xi \tag{N.207g}$$

$$\gamma = \frac{C_0 \beta}{s\pi \sqrt{D_x D_y}}[1 - \exp(t_s s)] \tag{N.207h}$$

$$\beta = [v^2/(4D_x) + \Theta]^{1/2} \tag{N.207i}$$

$$\Theta = s + \lambda + \frac{\rho_b \alpha k_d s}{\theta(s + \alpha)} \tag{N.207j}$$

Evaluation m-file:

```
function c=model_207(s,pv,x,y)

theta=pv(1);
rhob=pv(2);
Dx=pv(3);
vx=pv(4);
Dy=pv(5);
lambda=pv(6);
alpha=pv(7);
kd=pv(8);
C0=pv(9);
ts=pv(10);
d=pv(11);
```

```
ss=s*ones(size(x));
nx=length(x);
ns=length(s);
if isinf(alpha)
    THETA=(ss+lambda)+rhob/theta*kd*ss;
else
    THETA=(ss+lambda)+rhob/theta*alpha*kd*ss./...
    (ss+alpha);
end
beta=sqrt(vx^2/4/Dx+THETA);
f=zeros(ns,nx);
term0=C0/pi/sqrt(Dx*1); %Dy in *1 location is merged...
    into integrant

ns=1;
if x(1) <= 0, ns=2; end
for n=ns:nx
    if x(n)<=0, x(n)=1e-3; end
    term1=x(n)*exp(vx*x(n)/2/Dx);
    term2=quadv(@(w)thefun(w,Dx,Dy,beta,x(n),y),-d,d);
    f(:,n) = term0 * term1 * term2 .* beta;
end
if ns==2
    if abs(y)<d
        f(:,1)=C0;
    else
        f(:,1)=1e-15;
    end
end
f(f==0)=1e-15;
c = f ./ ss;

function f2=thefun(w,Dx,Dy,beta,x,y)
%slug=1-exp(-s*ts);
%slug is treated by superposition in solver
q=sqrt(x^2/Dx+(y-w)^2/Dy);
q0=sqrt(x^2/Dx*Dy +(y-w)^2); % Dy is in
z=beta*q;
f2=besselk(1,z)/q0;

end
```

N.18 Model 208

Governing equations:

$$\frac{\partial C}{\partial t} + \frac{\rho_b}{\theta}\frac{\partial S}{\partial t} = D_x\frac{\partial^2 C}{\partial x^2} - v\frac{\partial C}{\partial x} + D_y\frac{\partial^2 C}{\partial y^2} - \lambda C \tag{N.208a}$$

$$\frac{\partial S}{\partial t} = \alpha(k_d C - S) \tag{N.208b}$$

Initial conditions:

$$C = M\delta(x, y), \ t = 0, -\infty < x, y \le \infty \tag{N.208c}$$

$$S = 0, \ t = 0, -\infty < x, y \le \infty \tag{N.208d}$$

Boundary conditions:

$$C = 0, \ x, y \to \pm\infty, \ t > 0 \tag{N.208e}$$

Solution:

$$\overline{C}(x, s) = \frac{M}{2\pi\sqrt{D_x D_y}} \exp\left(\frac{vx}{2D_x}\right) K_0(\sqrt{r\omega}) \tag{N.208f}$$

$$r = \frac{x^2}{D_x} + \frac{y^2}{D_y} \tag{N.208g}$$

$$\omega = v^2/(4D_x) + \Theta \tag{N.208h}$$

$$\Theta = s + \lambda + \frac{\rho_b \alpha k_d s}{\theta(s + \alpha)} \tag{N.208i}$$

Evaluation m-file:

```
function c=model_208(s,pv,x,y)

theta=pv(1);
rhob=pv(2);
Dx=pv(3);
vx=pv(4);
Dy=pv(5);
lambda=pv(6);
alpha=pv(7);
kd=pv(8);
m0=pv(9);

x(x==0)=1e-4; % if x,y=0, singular

ss=s*ones(size(x));
```

```
X=ones(size(s))*x;
if isinf(alpha)
    THETA=(ss+lambda)+rhob/theta*kd*ss;
else
    THETA=(ss+lambda)+rhob/theta*alpha*kd*ss./...
    (ss+alpha);
end

w=m0/2/pi/sqrt(Dx*Dy);
r=X.^2/Dx + y^2/Dy;
q=vx^2/4/Dx + THETA;
z=sqrt(r.*q);
f = besselk(0,z);
c= w * exp(vx*X/2/Dx) .* f;

end
```

N.19   Model 301

Governing equations:

$$\frac{\partial C}{\partial t} + \frac{\rho_b}{\theta}\frac{\partial S}{\partial t} = D_x\frac{\partial^2 C}{\partial x^2} - v\frac{\partial C}{\partial x} + D_y\frac{\partial^2 C}{\partial y^2} + D_z\frac{\partial^2 C}{\partial z^2} - \lambda C \qquad \text{(N.301a)}$$

$$\frac{\partial S}{\partial t} = \alpha(k_d C - S) \qquad \text{(N.301b)}$$

Initial conditions:

$$C = S = 0, \ t = 0, \ 0 \le x \le l, \ 0 \le y \le w, \ 0 \le z \le h \qquad \text{(N.301c)}$$

Boundary conditions:

$$C = \begin{cases} C_0 H(t_s - t), & y_b \le y \le y_t, z_b \le z \le z_t \\ 0, & y_b > y > y_t, z_b > z > z_t \end{cases}, \ x = 0, \ t > 0 \qquad \text{(N.301d)}$$

$$C = 0, \ x = l, \ 0 \le y \le w, \ 0 \le z \le h, \ t > 0 \qquad \text{(N.301e)}$$

$$C = 0, \ y = 0, w, \ 0 \le x \le l, \ 0 \le z \le h, \ t > 0 \qquad \text{(N.301f)}$$

$$C = 0, \ z = 0, h, \ 0 \le x \le l, \ 0 \le y \le w, \ t > 0 \qquad \text{(N.301g)}$$

Solution:

$$\overline{C}(x,s) = \frac{4}{wh}\sum_{n=1}^{\infty}\sum_{m=1}^{\infty}\hat{\overline{C}}(x,n,m)\sin\left(\frac{n\pi y}{w}\right)\sin\left(\frac{m\pi z}{h}\right) \qquad \text{(N.301h)}$$

$$\hat{\bar{C}}(x, n, m) = \frac{1}{s}C_0[1 - \exp(-t_s s)]\gamma_n\gamma_m \frac{\exp(r_2 l + r_1 x) - \exp(r_1 l + r_2 x)}{\exp(r_2 l) - \exp(r_1 l)}$$

(N.301i)

$$\gamma_n = \frac{w}{n\pi}[\cos(n\pi y_t/w) - \cos(n\pi y_b/w)]$$

(N.301j)

$$\gamma_m = \frac{h}{m\pi}[\cos(m\pi z_t/h) - \cos(m\pi z_b/h)]$$

(N.301k)

$$r_{1,2} = \frac{1}{2D_x}(v \pm \{v^2 + 4D_x[D_y(n\pi/w)^2 + D_z(m\pi/h)^2 + \Theta]\}^{1/2})$$

(N.301l)

$$\Theta = s + \lambda + \frac{\rho_b \alpha k_d s}{\theta(s + \alpha)}$$

(N.301m)

Evaluation m-file:

```
function c=model_301(s,pv,xd,yd,zd,xyz)

theta=pv(1);
rhob=pv(2);
Dx=pv(3);
vx=pv(4);
Dy=pv(5);
Dz=pv(6);
lambda=pv(7);
alpha=pv(8);
kd=pv(9);
C0=pv(10);
ts=pv(11);
l=pv(12);
w=pv(13);
h=pv(14);
yb=pv(15);
yt=pv(16);
zb=pv(17);
zt=pv(18);

if xyz==1 %x-y plane
    x=xd; y=yd; z=zd;
elseif xyz==2 %x-z plane
    x=xd; z=yd; y=zd;
elseif xyz==3 %y-z plane
    y=xd; z=yd; x=zd;
```

```
end

if isinf(alpha)
    THETA=(s+lambda)+rhob/theta*kd*s;
else
    THETA=(s+lambda)+rhob/theta*alpha*kd*s./(s+alpha);
end

n=60;
m=60;
N=(1:n);
M=(1:m);
ss=s*ones(size(x));

gns=w/pi./N.*(cos(N*pi*yt/w)-cos(N*pi*yb/w))....
    *sin(N*pi*y/w);
gms=h/pi./M.*(cos(M*pi*zt/h)-cos(M*pi*zb/h))....
    *sin(M*pi*z/h);

NN=N'*ones(1,m);
MM=ones(n,1)*M;
NT=4*Dx*Dy*(NN*pi/w).^2;
MT=4*Dx*Dz*(MM*pi/h).^2;
f=zeros(length(s),length(x));

for i=1:length(s) %loop for s(i)
    for k=1:length(x) %loop for x(k)
        Delt=sqrt(vx*vx+NT+MT+4*Dx*THETA(i));
        Func=(exp(-Delt*l/Dx+x(k)*(vx+Delt)/2/Dx)- ...
            exp(x(k)*(vx-Delt)/2/Dx))./...
            (exp(-Delt*l/Dx)-1);
f(i,k) = gns * Func * gms';
end

end
f(f==0)=1e-15;
%slug=1-exp(-s*ts);
%slug is treated by superposition solutions in solver
c= 4/w/h*C0 * f./ss;

end
```

N.20 Model 302

Governing equations:

$$\frac{\partial C}{\partial t} + \frac{\rho_b}{\theta}\frac{\partial S}{\partial t} = D_x\frac{\partial^2 C}{\partial x^2} - v\frac{\partial C}{\partial x} + D_y\frac{\partial^2 C}{\partial y^2} + D_z\frac{\partial^2 C}{\partial z^2} - \lambda C \tag{N.302a}$$

$$\frac{\partial S}{\partial t} = \alpha(k_d C - S) \tag{N.302b}$$

Initial conditions:

$$C = S = 0,\ t = 0,\ 0 \le x \le l,\ 0 \le y \le w,\ 0 \le z \le h \tag{N.302c}$$

Boundary conditions:

$$C = \begin{cases} C_0 H(t_s - t), & y_b \le y \le y_t,\ z_b \le z \le z_t \\ 0, & y_b > y > y_t,\ z_b > z > z_t \end{cases},\ x = 0,\ t > 0 \tag{N.302d}$$

$$C = 0,\ x = l,\ 0 \le y \le w,\ 0 \le z \le h,\ t > 0 \tag{N.302e}$$

$$\partial C/\partial y = 0,\ y = 0, w,\ 0 \le x \le l,\ 0 \le z \le h,\ t > 0 \tag{N.302f}$$

$$\partial C/\partial z = 0,\ z = 0, h,\ 0 \le x \le l,\ 0 \le y \le w,\ t > 0 \tag{N.302g}$$

Solution:

$$\overline{C}(x,s) = \frac{4}{wh}\sum_{n=1}^{\infty}\sum_{m=1}^{\infty}\hat{\overline{C}}(x,n,m)\cos\left(\frac{n\pi y}{w}\right)\cos\left(\frac{m\pi z}{h}\right) \tag{N.302h}$$

$$\hat{\overline{C}}(x,n,m) = \frac{1}{s}C_0[1 - \exp(-t_s s)]\gamma_n\gamma_m\frac{\exp(r_2 l + r_1 x) - \exp(r_1 l + r_2 x)}{\exp(r_2 l) - \exp(r_1 l)} \tag{N.302i}$$

$$\gamma_n = \frac{w}{n\pi}[\sin(n\pi y_t/w) - \sin(n\pi y_b/w)]\delta_n \tag{N.302j}$$

$$\gamma_m = \frac{h}{m\pi}[\sin(m\pi z_t/h) - \sin(m\pi z_b/h)]\delta_m \tag{N.302k}$$

$$r_{1,2} = \frac{1}{2D_x}(v \pm \{v^2 + 4D_x[D_y(n\pi/w)^2 + D_z(m\pi/h)^2 + \Theta]\}^{1/2}) \tag{N.302l}$$

$$\Theta = s + \lambda + \frac{\rho_b\alpha k_d s}{\theta(s + \alpha)} \tag{N.302m}$$

$$\delta_i = \begin{cases} 1/2, & i = 0 \\ 1, & i > 0 \end{cases} \tag{N.302n}$$

Evaluation m-file:

```
function c=model_302(s,pv,xd,yd,zd,xyz)

theta=pv(1);
rhob=pv(2);
Dx=pv(3);
vx=pv(4);
Dy=pv(5);
Dz=pv(6);
lambda=pv(7);
alpha=pv(8);
kd=pv(9);
C0=pv(10);
ts=pv(11);
l=pv(12);
w=pv(13);
h=pv(14);
yb=pv(15);
yt=pv(16);
zb=pv(17);
zt=pv(18);

if xyz==1 %x-y plane
    x=xd; y=yd; z=zd;
elseif xyz==2 %x-z plane
    x=xd; z=yd; y=zd;
elseif xyz==3 %y-z plane
    y=xd; z=yd; x=zd;
end

if isinf(alpha)
    THETA=(s+lambda)+rhob/theta*kd*s;
else
    THETA=(s+lambda)+rhob/theta*alpha*kd*s./(s+alpha);
end

n=60;
m=60;
N=(1:n);
M=(1:m);
ss=s*ones(size(x));
```

```
gns=[(yt-yb)/2,...
     w/pi./N.*(sin(N*pi*yt/w)-sin(N*pi*yb/w))....
     *cos(N*pi*y/w)];
gms=[(zt-zb)/2,...
     h/pi./M.*(sin(M*pi*zt/h)-sin(M*pi*zb/h))....
     *cos(M*pi*z/h)];

NN=[0,N]'*ones(1,m+1);
MM=ones(n+1,1)*[0,M];
NT=4*Dx*Dy*(NN*pi/w).^2;
MT=4*Dx*Dz*(MM*pi/h).^2;
f=zeros(length(s),length(x));

lens=length(s);
lenx=length(x);
for i=1:lens %loop for s(i)
    for k=1:lenx %loop for x(k)
        Delt=sqrt(vx*vx+NT+MT+4*Dx*THETA(i));
        Func=(exp(-Delt*l/Dx+x(k)*(vx+Delt)/2/Dx)- ...
             exp(x(k)*(vx-Delt)/2/Dx))./...
             (exp(-Delt*l/Dx)-1);
        f(i,k) = gns * Func * gms';
    end
end

f(f==0)=1e-15;
%slug=1-exp(-s*ts);
%slug is treated by superposition solutions in solver
c= 4/w/h*C0 * f./ss;

end
```

N.21 Model 303

Governing equations:

$$\frac{\partial C}{\partial t} + \frac{\rho_b}{\theta}\frac{\partial S}{\partial t} = D_x\frac{\partial^2 C}{\partial x^2} - v\frac{\partial C}{\partial x} + D_y\frac{\partial^2 C}{\partial y^2} + D_z\frac{\partial^2 C}{\partial z^2} - \lambda C \qquad \text{(N.303a)}$$

$$\frac{\partial S}{\partial t} = \alpha(k_d C - S) \qquad \text{(N.303b)}$$

Initial conditions:

$$C = S = 0, \ t = 0, \ 0 \le x \le l, \ 0 \le y \le w, \ 0 \le z \le h \qquad \text{(N.303c)}$$

Boundary conditions:

$$C = \begin{cases} C_0 H(t_s - t), & y_b \le y \le y_t,\, z_b \le z \le z_t \\ 0, & y_b > y > y_t,\, z_b > z > z_t \end{cases},\quad x = 0,\, t > 0 \qquad \text{(N.303d)}$$

$$\partial C/\partial x = 0,\ x = l,\ 0 \le y \le w,\ 0 \le z \le h,\ t > 0 \qquad \text{(N.303e)}$$

$$\partial C/\partial y = 0,\ y = 0, w,\ 0 \le x \le l,\ 0 \le z \le h,\ t > 0 \qquad \text{(N.303f)}$$

$$\partial C/\partial z = 0,\ z = 0, h,\ 0 \le x \le l,\ 0 \le y \le w,\ t > 0 \qquad \text{(N.303g)}$$

Solution:

$$\overline{C}(x, s) = \frac{4}{wh} \sum_{n=1}^{\infty} \sum_{m=1}^{\infty} \hat{\overline{C}}(x, n, m) \cos\left(\frac{n\pi y}{w}\right) \cos\left(\frac{m\pi z}{h}\right) \qquad \text{(N.303h)}$$

$$\hat{\overline{C}}(x, n, m) = \frac{1}{s} C_0 [1 - \exp(-t_s s)] \gamma_n \gamma_m$$
$$\times\ \frac{r_2 \exp(r_2 l + r_1 x) - r_1 \exp(r_1 l + r_2 x)}{r_2 \exp(r_2 l) - r_1 \exp(r_1 l)} \qquad \text{(N.303i)}$$

$$\gamma_n = \frac{w}{n\pi} [\sin(n\pi y_t/w) - \sin(n\pi y_b/w)] \delta_n \qquad \text{(N.303j)}$$

$$\gamma_m = \frac{h}{m\pi} [\sin(m\pi z_t/h) - \sin(m\pi z_b/h)] \delta_m \qquad \text{(N.303k)}$$

$$r_{1,2} = \frac{1}{2D_x} (v \pm \{v^2 + 4D_x [D_y(n\pi/w)^2 + D_z(m\pi/h)^2 + \Theta]\}^{1/2}) \qquad \text{(N.303l)}$$

$$\Theta = s + \lambda + \frac{\rho_b \alpha k_d s}{\theta(s + \alpha)} \qquad \text{(N.303m)}$$

$$\delta_i = \begin{cases} 1/2, & i = 0 \\ 1, & i > 0 \end{cases} \qquad \text{(N.303n)}$$

Evaluation m-file:

```
function c=model_303(s,pv,xd,yd,zd,xyz)

theta=pv(1);
rhob=pv(2);
Dx=pv(3);
vx=pv(4);
Dy=pv(5);
Dz=pv(6);
lambda=pv(7);
alpha=pv(8);
kd=pv(9);
```

```
C0=pv(10);
ts=pv(11);
l=pv(12);
w=pv(13);
h=pv(14);
yb=pv(15);
yt=pv(16);
zb=pv(17);
zt=pv(18);

if xyz==1 %x-y plane
    x=xd; y=yd; z=zd;
elseif xyz==2 %x-z plane
    x=xd; z=yd; y=zd;
elseif xyz==3 %y-z plane
    y=xd; z=yd; x=zd;
end

if isinf(alpha)
    THETA=(s+lambda)+rhob/theta*kd*s;
else
    THETA=(s+lambda)+rhob/theta*alpha*kd*s./(s+alpha);
end

n=60;
m=60;
N=(1:n);
M=(1:m);
ss=s*ones(size(x));

gns=[(yt-yb)/2,...
     w/pi./N.*(sin(N*pi*yt/w)-sin(N*pi*yb/w))....
     *cos(N*pi*y/w)];
gms=[(zt-zb)/2,...
     h/pi./M.*(sin(M*pi*zt/h)-sin(M*pi*zb/h))....
     *cos(M*pi*z/h)];

NN=[0,N]'*ones(1,m+1);
MM=ones(n+1,1)*[0,M];
NT=4*Dx*Dy*(NN*pi/w).^2;
MT=4*Dx*Dz*(MM*pi/h).^2;
f=zeros(length(s),length(x));
```

```
lens=length(s);
lenx=length(x);
for i=1:lens %loop for s(i)
    for k=1:lenx %loop for x(k)
        Delt=sqrt(vx*vx+NT+MT+4*Dx*THETA(i));
        r1=(vx+Delt)/2/Dx;
        r2=(vx-Delt)/2/Dx;
        Func=(r2.*exp(-Delt*l/Dx+x(k)...
         *(vx+Delt)/2/Dx)-...
         r1.*exp(x(k)*(vx-Delt)/2/Dx))./...
         (r2.*exp(-Delt*l/Dx)-r1);
        f(i,k) = gns * Func * gms';
    end
end

f(f==0)=1e-15;
%slug=1-exp(-s*ts);
%slug is treated by superposition solutions in solver
c= 4/w/h*C0 * f./ss;

end
```

N.22 Model 304

Governing equations:

$$\frac{\partial C}{\partial t} + \frac{\rho_b}{\theta}\frac{\partial S}{\partial t} = D_x\frac{\partial^2 C}{\partial x^2} - v\frac{\partial C}{\partial x} + D_y\frac{\partial^2 C}{\partial y^2} + D_z\frac{\partial^2 C}{\partial z^2} - \lambda C \qquad \text{(N.304a)}$$

$$\frac{\partial S}{\partial t} = \alpha(k_d C - S) \qquad \text{(N.304b)}$$

Initial conditions:

$$C = S = 0, \ t = 0, \ 0 \le x \le l, \ 0 \le y \le w, \ 0 \le z \le h \qquad \text{(N.304c)}$$

Boundary conditions:

$$C = \begin{cases} C_0 H(t_s - t), & y_b \le y \le y_t, z_b \le z \le z_t \\ 0, & y_b > y > y_t, z_b > z > z_t \end{cases}, \ x = 0, \ t > 0 \qquad \text{(N.304d)}$$

$$C = 0, \ x \to \infty, \ 0 \le y \le w, \ 0 \le z \le h, \ t > 0 \qquad \text{(N.304e)}$$

$$C = 0, \ y = 0, w, \ 0 \le x \le l, \ 0 \le z \le h, \ t > 0 \qquad \text{(N.304f)}$$

$$C = 0, \ z = 0, h, \ 0 \le x \le l, \ 0 \le y \le w, \ t > 0 \qquad \text{(N.304g)}$$

Solution:

$$\overline{C}(x, s) = \frac{4}{wh} \sum_{n=1}^{\infty} \sum_{m=1}^{\infty} \hat{\overline{C}}(x, n, m) \sin\left(\frac{n\pi y}{w}\right) \sin\left(\frac{m\pi z}{h}\right) \qquad \text{(N.304h)}$$

$$\hat{\overline{C}}(x, n, m) = \frac{1}{s} C_0 [1 - \exp(-t_s s)] \gamma_n \gamma_m \exp(rx) \qquad \text{(N.304i)}$$

$$\gamma_n = \frac{w}{n\pi} [\cos(n\pi y_t/w) - \cos(n\pi y_b/w)] \qquad \text{(N.304j)}$$

$$\gamma_m = \frac{h}{m\pi} [\cos(m\pi z_t/h) - \cos(m\pi z_b/h)] \qquad \text{(N.304k)}$$

$$r = \frac{1}{2D_x}(v - \{v^2 + 4D_x[D_y(n\pi/w)^2 + D_z(m\pi/h)^2 + \Theta]\}^{1/2}) \qquad \text{(N.304l)}$$

$$\Theta = s + \lambda + \frac{\rho_b \alpha k_d s}{\theta(s + \alpha)} \qquad \text{(N.304m)}$$

Evaluation m-file:

```
function c=model_304(s,pv,xd,yd,zd,xyz)

theta=pv(1);
rhob=pv(2);
Dx=pv(3);
vx=pv(4);
Dy=pv(5);
Dz=pv(6);
lambda=pv(7);
alpha=pv(8);
kd=pv(9);
C0=pv(10);
ts=pv(11);
w=pv(12);
h=pv(13);
yb=pv(14);
yt=pv(15);
zb=pv(16);
zt=pv(17);

if xyz==1 %x-y plane
    x=xd; y=yd; z=zd;
elseif xyz==2 %x-z plane
    x=xd; z=yd; y=zd;
elseif xyz==3 %y-z plane
    y=xd; z=yd; x=zd;
end
```

```
if isinf(alpha)
    THETA=(s+lambda)+rhob/theta*kd*s;
else
    THETA=(s+lambda)+rhob/theta*alpha*kd*s./(s+alpha);
end

n=60;
m=60;
N=(1:n);
M=(1:m);
ss=s*ones(size(x));

gns=w/pi./N.*(cos(N*pi*yt/w)-cos(N*pi*yb/w))....
    *sin(N*pi*y/w);
gms=h/pi./M.*(cos(M*pi*zt/h)-cos(M*pi*zb/h))....
    *sin(M*pi*z/h);

NN=N'*ones(1,m);
MM=ones(n,1)*M;
NT=4*Dx*Dy*(NN*pi/w).^2;
MT=4*Dx*Dz*(MM*pi/h).^2;
f=zeros(length(s),length(x));

for i=1:length(s) %loop for s(i)
    for k=1:length(x) %loop for x(k)
        Delt=sqrt(vx*vx+NT+MT+4*Dx*THETA(i));
        Func=exp(x(k)*(vx-Delt)/2/Dx);
        f(i,k) = gns * Func * gms';
    end
end
f(f==0)=1e-15;
%slug=1-exp(-s*ts);
%slug is treated by superposition solutions in solver
c= 4/w/h*C0 * f./ss;

end
```

N.23   Model 305

Governing equations:

$$\frac{\partial C}{\partial t}+\frac{\rho_b}{\theta}\frac{\partial S}{\partial t}=D_x\frac{\partial^2 C}{\partial x^2}-v\frac{\partial C}{\partial x}+D_y\frac{\partial^2 C}{\partial y^2}+D_z\frac{\partial^2 C}{\partial z^2}-\lambda C \qquad \text{(N.305a)}$$

$$\frac{\partial S}{\partial t} = \alpha(k_d C - S) \tag{N.305b}$$

Initial conditions:

$$C = S = 0, \ t = 0, \ 0 \le x \le l, \ 0 \le y \le w, \ 0 \le z \le h \tag{N.305c}$$

Boundary conditions:

$$C = \begin{cases} C_0 H(t_s - t), & y_b \le y \le y_t, \ z_b \le z \le z_t \\ 0, & y_b > y > y_t, \ z_b > z > z_t \end{cases}, \ x = 0, \ t > 0 \tag{N.305d}$$

$$C = 0, \ x \to \infty, \ 0 \le y \le w, \ 0 \le z \le h, \ t > 0 \tag{N.305e}$$

$$\partial C/\partial y = 0, \ y = 0, w, \ 0 \le x \le l, \ 0 \le z \le h, \ t > 0 \tag{N.305f}$$

$$\partial C/\partial z = 0, \ z = 0, h, \ 0 \le x \le l, \ 0 \le y \le w, \ t > 0 \tag{N.305g}$$

Solution:

$$\overline{C}(x, s) = \frac{4}{wh} \sum_{n=1}^{\infty} \sum_{m=1}^{\infty} \hat{\overline{C}}(x, n, m) \cos\left(\frac{n\pi y}{w}\right) \cos\left(\frac{m\pi z}{h}\right) \tag{N.305h}$$

$$\hat{\overline{C}}(x, n, m) = \frac{1}{s} C_0 [1 - \exp(-t_s s)] \gamma_n \gamma_m \exp(rx) \tag{N.305i}$$

$$\gamma_n = \frac{w}{n\pi} [\sin(n\pi y_t/w) - \sin(n\pi y_b/w)] \delta_n \tag{N.305j}$$

$$\gamma_m = \frac{h}{m\pi} [\sin(m\pi z_t/h) - \sin(m\pi z_b/h)] \delta_m \tag{N.305k}$$

$$r = \frac{1}{2D_x} (v - \{v^2 + 4D_x [D_y(n\pi/w)^2 + D_z(m\pi/h)^2 + \Theta]\}^{1/2}) \tag{N.305l}$$

$$\Theta = s + \lambda + \frac{\rho_b \alpha k_d s}{\theta(s + \alpha)} \tag{N.305m}$$

$$\delta_i = \begin{cases} 1/2, & i = 0 \\ 1, & i > 0 \end{cases} \tag{N.305n}$$

Evaluation m-file:

```
function c=model_305(s,pv,xd,yd,zd,xyz)

theta=pv(1);
rhob=pv(2);
Dx=pv(3);
vx=pv(4);
Dy=pv(5);
Dz=pv(6);
lambda=pv(7);
```

```
alpha=pv(8);
kd=pv(9);
C0=pv(10);
ts=pv(11);
w=pv(12);
h=pv(13);
yb=pv(14);
yt=pv(15);
zb=pv(16);
zt=pv(17);

if xyz==1 %x-y plane
    x=xd; y=yd; z=zd;
elseif xyz==2 %x-z plane
    x=xd; z=yd; y=zd;
elseif xyz==3 %y-z plane
    y=xd; z=yd; x=zd;
end

if isinf(alpha)
    THETA=(s+lambda)+rhob/theta*kd*s;
else
    THETA=(s+lambda)+rhob/theta*alpha*kd*s./(s+alpha);
end

n=60;
m=60;
N=(1:n);
M=(1:m);
ss=s*ones(size(x));

gns=[(yt-yb)/2,...
    w/pi./N.*(sin(N*pi*yt/w)-sin(N*pi*yb/w))....
    *cos(N*pi*y/w)];
gms=[(zt-zb)/2,...
    h/pi./M.*(sin(M*pi*zt/h)-sin(M*pi*zb/h))....
    *cos(M*pi*z/h)];

NN=[0,N]'*ones(1,m+1);
MM=ones(n+1,1)*[0,M];
NT=4*Dx*Dy*(NN*pi/w).^2;
MT=4*Dx*Dz*(MM*pi/h).^2;
f=zeros(length(s),length(x));
```

```
lens=length(s);
lenx=length(x);
for i=1:lens %loop for s(i)
    for k=1:lenx %loop for x(k)
        Delt=sqrt(vx*vx+NT+MT+4*Dx*THETA(i));
        Func=exp(x(k)*(vx-Delt)/2/Dx);
        f(i,k) = gns * Func * gms';
    end
end

f(f==0)=1e-15;
%slug=1-exp(-s*ts);
%slug is treated by superposition solutions in solver
c= 4/w/h*C0 * f./ss;

end
```

N.24   Model 306

Governing equations:

$$\frac{\partial C}{\partial t} + \frac{\rho_b}{\theta}\frac{\partial S}{\partial t} = D_x\frac{\partial^2 C}{\partial x^2} - v\frac{\partial C}{\partial x} + D_y\frac{\partial^2 C}{\partial y^2} + D_z\frac{\partial^2 C}{\partial z^2} - \lambda C \qquad (N.306a)$$

$$\frac{\partial S}{\partial t} = \alpha(k_d C - S) \qquad (N.306b)$$

Initial conditions:

$$C = M\delta(x, y), \ t = 0, -\infty < x, y, z \le \infty \qquad (N.306c)$$

$$S = 0, \ t = 0, -\infty < x, y, z \le \infty \qquad (N.306d)$$

Boundary conditions,:

$$C = 0, \ x, y, z \rightarrow \pm\infty, \ t > 0 \qquad (N.306e)$$

Solution:

$$\overline{C}(x, s) = \frac{M}{4\pi\sqrt{D_x D_y D_z}}\frac{1}{G}\exp\left(\frac{vx}{2D_x} - G\Omega(s)\right) \qquad (N.306f)$$

$$G = (x^2/D_x + y^2/D_y + z^2/D_z)^{1/2} \qquad (N.306g)$$

$$\Omega(s) = [v^2/(4D_x) + \Theta]^{1/2} \qquad (N.306h)$$

$$\Theta = s + \lambda + \frac{\rho_b \alpha k_d s}{\theta(s + \alpha)} \qquad (N.306i)$$

Evaluation m-file:

```
function c=model_306(s,pv,xd,yd,zd,xyz)

theta=pv(1);
rhob=pv(2);
Dx=pv(3);
vx=pv(4);
Dy=pv(5);
Dz=pv(6);
lambda=pv(7);
alpha=pv(8);
kd=pv(9);
m0=pv(10);

if xyz==1 %x-y plane
    x=xd; y=yd; z=zd;
elseif xyz==2 %x-z plane
    x=xd; z=yd; y=zd;
elseif xyz==3 %y-z plane
    y=xd; z=yd; x=zd;
end

if isinf(alpha)
    THETA=(s+lambda)+rhob/theta*kd*s;
else
    THETA=(s+lambda)+rhob/theta*alpha*kd*s./(s+alpha);
end

OMEGA=sqrt(vx*vx/4/Dx+THETA);
G=sqrt(x*x/Dx+y*y/Dy+z*z/Dz);

if G<=0;
    f=m0/4/pi/sqrt(Dx*Dy*Dz)*(-OMEGA);
else
    f=m0/4/pi/sqrt(Dx*Dy*Dz)/G...
    * exp(vx*x/2/Dx-G*OMEGA);
end

c=f;

end
```

N.25   Model 401

Governing equations:

$$\frac{\partial C}{\partial t} + \frac{\rho_b}{\theta}\frac{\partial S}{\partial t} = \frac{a_L q}{r}\frac{\partial^2 C}{\partial r^2} - \frac{q}{r}\frac{\partial C}{\partial r} - \lambda C \tag{N.401a}$$

$$\frac{\partial S}{\partial t} = \alpha(k_d C - S) \tag{N.401b}$$

Initial conditions:

$$C = S = 0,\ t = 0,\ r_w \le r < \infty \tag{N.401c}$$

Boundary conditions:

$$C = C_0 H(t_s - t),\ r = r_w,\ t > 0 \tag{N.401d}$$

$$C = 0,\ r \to \infty,\ t > 0 \tag{N.401e}$$

Solution:

$$\overline{C}(r,s) = \frac{1}{s}C_0[1 - \exp(-t_s s)]\exp\left(\frac{r - r_w}{2a_L}\right)\frac{Ai(Y)}{Ai(Y_w)} \tag{N.401f}$$

$$Y = \left(r + \frac{1}{4a_L z}\right)(z/a_L)^{1/3} \tag{N.401g}$$

$$Y_w = \left(r_w + \frac{1}{4a_L z}\right)(z/a_L)^{1/3} \tag{N.401h}$$

$$z = \Theta/q \tag{N.401i}$$

$$\Theta = s + \lambda + \frac{\rho_b \alpha k_d s}{\theta(s + \alpha)} \tag{N.401j}$$

$$q = \frac{Q_w}{2\pi b\theta} \tag{N.401k}$$

Evaluation m-file:

```
function c=model_401(s,pv,r,sflag)

theta=pv(1);
rhob=pv(2);
al=pv(3);
Q=pv(4);
rw=pv(5);
lambda=pv(6);
alpha=pv(7);
kd=pv(8);
C0=pv(9);
```

```
b=pv(10);
ts=pv(11);

q=Q/(2*pi*b*theta);

S=s*ones(size(r));
X=ones(size(s))*r;
if isinf(alpha)
    THETA=(S+lambda)+rhob/theta*kd*S;
else
    THETA=(S+lambda)+rhob/theta*alpha*kd*S./(S+alpha);
end

z=THETA/q;
u=1.0/4/al ./ z;
v=(z/al) .^ (1./3.);
Y=(X+u).*v;
Yw=(rw+u).*v;

%slug=1-exp(-s*ts);
%slug is treated by superposition solutions in solver

c=C0 ./ S .* exp((X-rw)/2/al) .* airy(Y)./airy(Yw);

end
```

N.26   Model 402

Governing equations:

$$\frac{\partial C}{\partial t} + \frac{\rho_b}{\theta}\frac{\partial S}{\partial t} = \frac{a_L q}{r}\frac{\partial^2 C}{\partial r^2} - \frac{q}{r}\frac{\partial C}{\partial r} - \lambda C \tag{N.402a}$$

$$\frac{\partial S}{\partial t} = \alpha(k_d C - S) \tag{N.402b}$$

Initial conditions:

$$C = S = 0, \ t = 0, \ r_w \leq r < \infty \tag{N.402c}$$

Boundary conditions:

$$-a_L\frac{\partial C}{\partial r} + C = C_0 H(t_s - t), \ r = r_w, \ t > 0 \tag{N.402d}$$

$$C = 0, \ r \to \infty, \ t > 0 \tag{N.402e}$$

Solution:

$$\overline{C}(r,s) = \frac{1}{s}C_0[1 - \exp(-t_s s)] \exp\left(\frac{r - r_w}{2a_L}\right) \frac{2Ai(Y)}{Ai(Y_w) - 2z^{1/3}a_L^{2/3}Ai'(Y_w)}$$

(N.402f)

$$Y = \left(r + \frac{1}{4a_L z}\right)(z/a_L)^{1/3}$$

(N.402g)

$$Y_w = \left(r_w + \frac{1}{4a_L z}\right)(z/a_L)^{1/3}$$

(N.402h)

$$z = \Theta/q$$

(N.402i)

$$\Theta = s + \lambda + \frac{\rho_b \alpha k_d s}{\theta(s + \alpha)}$$

(N.402j)

$$q = \frac{Q_w}{2\pi b\theta}$$

(N.402k)

Evaluation m-file:

```
function c=model_402(s,pv,r,sflag)

theta=pv(1);
rhob=pv(2);
al=pv(3);
Q=pv(4);
rw=pv(5);
lambda=pv(6);
alpha=pv(7);
kd=pv(8);
C0=pv(9);
b=pv(10);
ts=pv(11);

q=Q/(2*pi*b*theta);
S=s*ones(size(r));
X=ones(size(s))*r;
if isinf(alpha)
    THETA=(S+lambda)+rhob/theta*kd*S;
else
    THETA=(S+lambda)+rhob/theta*alpha*kd*S./(S+alpha);
end

z=THETA/q;
u=1.0/4/al ./ z;
v=(z/al) .^ (1./3.);
```

```
Y=(X+u).*v;
Yw=(rw+u).*v;

%slug=1-exp(-s*ts);
%slug is treated by superposition solutions in solver

w=0.5*airy(Yw)-v*al .* airy(1,Yw);
c=C0 ./ S .* exp((X-rw)/2/al) .* airy(Y)./w;

end
```

N.27   Model 403

Governing equations:

$$\frac{\partial C}{\partial t} + \frac{\rho_b}{\theta}\frac{\partial S}{\partial t} = \frac{a_L q}{r}\frac{\partial^2 C}{\partial r^2} - \frac{q}{r}\frac{\partial C}{\partial r} - \lambda C \tag{N.403a}$$

$$\frac{\partial S}{\partial t} = \alpha(k_d C - S) \tag{N.403b}$$

Initial conditions:

$$C = C_0,\ S = k_d C_0,\ t = 0,\ r_w \le r \le r_b \tag{N.403c}$$

Boundary conditions:

$$\partial C/\partial r = 0,\ r = r_w,\ t > 0 \tag{N.403d}$$

$$C = 0,\ r = r_b,\ t > 0 \tag{N.403e}$$

Solution:

$$\overline{C}(r,s) = \frac{s + \alpha + \rho_b k_d \alpha/\theta}{(s+\alpha)\Theta} C_0\{1 - \exp[(r_b - r)/(2a_L)]\}$$
$$\times \frac{Ai(Y)U - Bi(Y)V}{Ai(Y_b)U - B(Y_b)V} \tag{N.403f}$$

$$U = Bi(Y_w) - \gamma Bi'(Y_w) \tag{N.403g}$$

$$V = Ai(Y_w) - \gamma Ai'(Y_w) \tag{N.403h}$$

$$Y = \left(r + \frac{1}{4a_L z}\right)(z/a_L)^{1/3} \tag{N.403i}$$

$$Y_w = \left(r_w + \frac{1}{4a_L z}\right)(z/a_L)^{1/3} \tag{N.403j}$$

$$Y_b = \left(r_b + \frac{1}{4a_L z}\right)(z/a_L)^{1/3} \qquad (N.403k)$$

$$z = \Theta/q \qquad (N.403l)$$

$$q = \frac{Q_w}{2\pi b\theta} \qquad (N.403m)$$

$$\gamma = 2a_L^{2/3}z^{1/3} \qquad (N.403n)$$

$$\Theta = s + \lambda + \frac{\rho_b \alpha k_d s}{\theta(s+\alpha)} \qquad (N.403o)$$

Evaluation m-file:

```
function c=model_403(s,pv,r,sflag)

theta=pv(1);
rhob=pv(2);
al=pv(3);
Q=pv(4);
rw=pv(5);
lambda=pv(6);
alpha=pv(7);
kd=pv(8);
C0=pv(9);
b=pv(10);
rb=pv(11);

q=Q/(2*pi*b*theta);

rlen=length(r);
if r(rlen)>=rb
    r(rlen)=rb*0.999;
end

S=s*ones(size(r));
X=ones(size(s))*r;
if isinf(alpha)
    THETA=(S+lambda)+rhob/theta*kd*S;
else
    THETA=(S+lambda)+rhob/theta*alpha*kd*S./(S+alpha);
end

z=THETA/q;
```

```
u=1.0/4/al ./ z;
v=(z/al) .^ (1./3.);
Y=(X+u).*v;
Yw=(rw+u).*v;
Yb=(rb+u).*v;
gam=2.*al^(2./3)*z.^(1./3);

bib=airy(2,Yb);

uu=(airy(2,Yw)-gam.*airy(3,Yw))./bib;
vv=airy(0,Yw)-gam.*airy(1,Yw);
w=exp((rb-X)/2./al);
b2b=airy(2,Y)./bib;
aa=airy(0,Y) .* uu - b2b .* vv;
bb=airy(0,Yb) .* uu - vv;

Cstar=C0 * (1+rhob/theta*kd*alpha./(S+alpha))./ THETA;
c=Cstar .* (1. - w .* aa./bb);

return
end
```

N.28 Model 404

Governing equations:

$$\frac{\partial C}{\partial t} + \frac{\rho_b}{\theta}\frac{\partial S}{\partial t} = \frac{a_L q}{r}\frac{\partial^2 C}{\partial r^2} - \frac{q}{r}\frac{\partial C}{\partial r} - \lambda C + \frac{2D_m}{b}\frac{\partial C_m}{\partial z}\bigg|_{z=0} \qquad \text{(N.404a)}$$

$$\frac{\partial S}{\partial t} = \alpha(k_d C - S) \qquad \text{(N.404b)}$$

$$\frac{\partial C_m}{\partial t} + \frac{\rho_m}{\theta_m}\frac{\partial S}{\partial t} = D_m\frac{\partial^2 C_m}{\partial r^2} - \lambda C_m \qquad \text{(N.404c)}$$

$$\frac{\partial S_m}{\partial t} = \alpha_m(k_m C_m - S_m) \qquad \text{(N.404d)}$$

Initial conditions:

$$C = S = C_m = S_m = 0,\ t = 0,\ r_w \leq r < \infty \qquad \text{(N.404e)}$$

Boundary conditions:

$$C = C_0 H(t_s - t),\ r = r_w,\ t > 0 \qquad \text{(N.404f)}$$

$$C_m = C,\ z = 0,\ r_w \leq r < \infty,\ t > 0 \qquad \text{(N.404g)}$$

$$C = 0, \ r \to \infty, \ t > 0 \tag{N.404h}$$

$$C_m = 0, \ z \to \infty, \ t > 0 \tag{N.404i}$$

Solution:

$$\overline{C}(r,s) = \frac{1}{s} C_0 [1 - \exp(-t_s s)] \exp\left(\frac{r - r_w}{2a_L}\right) \frac{Ai(Y)}{Ai(Y_w)} \tag{N.404j}$$

$$Y = \left(r + \frac{1}{4a_L z}\right)(z/a_L)^{1/3} \tag{N.404k}$$

$$Y_w = \left(r_w + \frac{1}{4a_L z}\right)(z/a_L)^{1/3} \tag{N.404l}$$

$$z = (\Theta + \Gamma)/q$$

$$\Theta = s + \lambda + \frac{\rho_b \alpha k_d s}{\theta(s + \alpha)} \tag{N.404m}$$

$$\Gamma = \frac{2}{b} \sqrt{D_m \Theta_m} \tag{N.404n}$$

$$\Theta_m = s + \lambda + \frac{\rho_m \alpha_m k_m s}{\theta_m(s + \alpha_m)} \tag{N.404o}$$

$$q = \frac{Q_w}{2\pi b \theta} \tag{N.404p}$$

Evaluation m-file:

```
function c=model_404(s,pv,r,sflag)

theta=pv(1);
rhob=pv(2);
al=pv(3);
Q=pv(4);
rw=pv(5);
lambda=pv(6);
alpha=pv(7);
kd=pv(8);
C0=pv(9);
thetam=pv(10);
rhom=pv(11);
Dm=pv(12);
alpham=pv(13);
km=pv(14);
b=pv(15);
ts=pv(16);
```

```
q=Q/(2*pi*b*theta);

S=s*ones(size(r));
X=ones(size(s))*r;
if isinf(alpha)
    THETA=(S+lambda)+rhob/theta*kd*S;
else
    THETA=(S+lambda)+rhob/theta*alpha*kd*S./(S+alpha);
end
if isinf(alpham)
    THETAM=(S+alpham)+rhom/thetam*km*S;
else
    THETAM=(S+alpham)+rhom/thetam*alpham*km*S./...
    (S+alpham);
end
GAMMA=2*sqrt(Dm*THETAM/b);

z=(THETA + GAMMA)/q;
u=1.0/4/al ./ z;
v=(z/al) .^ (1./3.);
Y=(X+u).*v;
Yw=(rw+u).*v;

%slug=1-exp(-s*ts);
%slug is treated by superposition solutions in solver

c=C0 ./ S .* exp((X-rw)/2/al) .* airy(Y)./airy(Yw);

return
end
```

N.29   Model 405

Governing equations:

$$\frac{\partial C}{\partial t} + \frac{\rho_b}{\theta}\frac{\partial S}{\partial t} = \frac{a_L q}{r}\frac{\partial^2 C}{\partial r^2} - \frac{q}{r}\frac{\partial C}{\partial r} - \lambda C + \frac{2D_m}{b}\left.\frac{\partial C_m}{\partial z}\right|_{z=0} \tag{N.405a}$$

$$\frac{\partial S}{\partial t} = \alpha(k_d C - S) \tag{N.405b}$$

$$\frac{\partial C_m}{\partial t} + \frac{\rho_m}{\theta_m}\frac{\partial S}{\partial t} = D_m\frac{\partial^2 C_m}{\partial r^2} - \lambda C_m \tag{N.405c}$$

$$\frac{\partial S_m}{\partial t} = \alpha_m(k_m C_m - S_m) \tag{N.405d}$$

Initial conditions:

$$C = S = C_m = S_m = 0, \ t = 0, \ r_w \le r < \infty \tag{N.405e}$$

Boundary conditions

$$-a_L \frac{\partial C}{\partial r} + C = C_0 H(t_s - t), \ r = r_w, \ t > 0 \tag{N.405f}$$

$$C_m = C, \ z = 0, \ r_w \le r < \infty, \ t > 0 \tag{N.405g}$$

$$C = 0, \ r \to \infty, \ t > 0 \tag{N.405h}$$

$$C_m = 0, \ z \to \infty, \ t > 0 \tag{N.405i}$$

Solution:

$$\overline{C}(r,s) = \frac{1}{s} C_0 [1 - \exp(-t_s s)] \exp\left(\frac{r - r_w}{2a_L}\right) \frac{2Ai(Y)}{Ai(Y_w) - 2z^{1/3} a_L^{2/3} Ai'(Y_w)} \tag{N.405j}$$

$$Y = \left(r + \frac{1}{4a_L z}\right)(z/a_L)^{1/3} \tag{N.405k}$$

$$Y_w = \left(r_w + \frac{1}{4a_L z}\right)(z/a_L)^{1/3} \tag{N.405l}$$

$$z = (\Theta + \Gamma)/q \tag{N.405m}$$

$$\Theta = s + \lambda + \frac{\rho_b \alpha k_d s}{\theta(s + \alpha)} \tag{N.405n}$$

$$\Gamma = \frac{2}{b}\sqrt{D_m \Theta_m} \tag{N.405o}$$

$$\Theta_m = s + \lambda + \frac{\rho_m \alpha_m k_m s}{\theta_m(s + \alpha_m)} \tag{N.405p}$$

$$q = \frac{Q_w}{2\pi b\theta} \tag{N.405q}$$

Evaluation m-file:

```
function c=model_405(s,pv,r,sflag)

theta=pv(1);
rhob=pv(2);
al=pv(3);
Q=pv(4);
rw=pv(5);
```

```
lambda=pv(6);
alpha=pv(7);
kd=pv(8);
C0=pv(9);
thetam=pv(10);
rhom=pv(11);
Dm=pv(12);
alpham=pv(13);
km=pv(14);
b=pv(15);
ts=pv(16);

q=Q/(2*pi*b*theta);

S=s*ones(size(r));
X=ones(size(s))*r;
if isinf(alpha)
    THETA=(S+lambda)+rhob/theta*kd*S;
else
    THETA=(S+lambda)+rhob/theta*alpha*kd*S./(S+alpha);
end
if isinf(alpham)
    THETAM=(S+alpham)+rhom/thetam*km*S;
else
    THETAM=(S+alpham)+rhom/thetam*alpham*km*S./...
    (S+alpham);
end
GAMMA=2*sqrt(Dm*THETAM/b);

z=(THETA + GAMMA)/q;
u=1.0/4/al ./ z;
v=(z/al) .^ (1./3.);
Y=(X+u).*v;
Yw=(rw+u).*v;

%slug=1-exp(-s*ts);
%slug is treated by superposition solutions in solver

w=0.5*airy(Yw)-v*al .* airy(1,Yw);
c=C0 ./ S .* exp((X-rw)/2/al) .* airy(Y)./w;

return
end
```

N.30 Model 406

Governing equations:

$$\frac{\partial C_m}{\partial T} = \frac{1}{X}\frac{\partial^2 C_m}{\partial X^2} + \frac{1}{X}\frac{\partial C_m}{\partial X} + \beta\frac{\partial C_{im}}{\partial T} \tag{N.406a}$$

$$C_{im} = \int_0^1 Z^{v-1} C_a dZ \tag{N.406b}$$

$$\frac{\partial C_a}{\partial T} = \frac{D_e}{Z^{v-1}}\frac{\partial}{\partial Z}\left(Z^{v-1}\frac{\partial C_a}{\partial Z}\right) \tag{N.406c}$$

Initial conditions:

$$C_m = C_{im} = C_a = C_0, \ T = 0, \ X_w \le X \le X_b \tag{N.406d}$$

Boundary conditions

$$\partial C_m/\partial X = 0, \ X = X_w, \ t > 0 \tag{N.406e}$$

$$\partial C_m/\partial X + C_m = 0, \ X = X_b, \ t > 0 \tag{N.406f}$$

$$C = 0, \ r \to \infty, \ t > 0 \tag{N.406g}$$

$$\partial C_a/\partial Z = 0, \ Z = 0, \ t > 0 \tag{N.406h}$$

$$C_a = C_m, \ Z = 1, \ t > 0 \tag{N.406i}$$

Solution:

$$\overline{C}_m(r,s) = \frac{1}{s}C_0[1 - \exp((X_b - X)/2)]\frac{Bi(y)G_A - Ai(y)G_B}{G_A H_B - G_B H_A} \tag{N.406j}$$

$$G_A = -\tfrac{1}{2}Ai(y_w) + \gamma^{1/3}Ai'(y_w) \tag{N.406k}$$

$$G_B = -\tfrac{1}{2}Bi(y_w) + \gamma^{1/3}Bi'(y_w) \tag{N.406l}$$

$$H_A = \tfrac{1}{2}Ai(y_b) + \gamma^{1/3}Ai'(y_b) \tag{N.406m}$$

$$H_B = \tfrac{1}{2}Bi(y_b) + \gamma^{1/3}Bi'(y_b) \tag{N.406n}$$

$$y = \gamma^{1/3}[X + 1/(4\gamma)] \tag{N.406o}$$

$$y_w = \gamma^{1/3}[X_w + 1/(4\gamma)] \tag{N.406p}$$

$$y_b = \gamma^{1/3}[X_b + 1/(4\gamma)] \tag{N.406q}$$

$$\gamma = s(1 + v\beta/\omega f_v) \tag{N.406r}$$

$$f_1 = \frac{\sinh(\omega)}{\cosh(\omega)}, \ f_2 = \frac{I_1(\omega)}{I_0(\omega)}, \ f_1 = \frac{i_1(\omega)}{i_0(\omega)} \tag{N.406s}$$

$$\omega = (s/D_e)^{1/2} \tag{N.406t}$$

$$X = r/a_L, \ T = Q_w t/(2\pi b\theta_m a_L^2 R_m), \ Z = z/a \tag{N.406u}$$

$$\beta = \theta_{im}R_{im}/(\theta_{im}R_{im}), \ D_e = D'_e a_L 2\pi b\theta_m R_m/(a^2 Q_w R_{im}) \tag{N.406v}$$

Evaluation m-file:

```
function c=model_406(s,pv,r,sflag)

beta0=pv(1);
nu=pv(2);
De=pv(3);
C0=pv(4);
Xw=pv(5);
Xb=pv(6);


[XX,ss]=meshgrid(r,s);

omega0=sqrt(ss/De);
if nu==1
    fnu=tanh(omega0);
elseif nu==2
    fnu=besseli(1,omega0,1)./besseli(0,omega0,1);
elseif nu==3
    fnu=cosh(omega0)./sinh(omega0)-1./omega0;
end
gamma0=ss.*(1+nu*beta0./omega0.*fnu);

g3=gamma0.^(1/3);

y =g3.*(XX+1/4./gamma0);
yw=g3.*(Xw+1/4./gamma0);
yb=g3.*(Xb+1/4./gamma0);

G_As=-0.5*airy(0,yw,1)+g3.*airy(1,yw,1);
G_Bs=-0.5*airy(2,yw,1)+g3.*airy(3,yw,1);
H_As=0.5*airy(0,yb,1)+g3.*airy(1,yb,1);
H_Bs=0.5*airy(2,yb,1)+g3.*airy(3,yb,1);

psAi=-2/3*y.^(3/2);
psBi=abs(2/3*real(y.^(3/2)));
```

```
psGAi=-2/3*yw.^(3/2);
psGBi=abs(2/3*real(yw.^(3/2)));
psHAi=-2/3*yb.^(3/2);
psHBi=abs(2/3*real(yb.^(3/2)));

w=exp(0.5*(Xb-XX)-psHBi);

aa=airy(2,y,1) .* G_As .* exp(psBi+psGAi) - ...
   airy(0,y,1) .* G_Bs .* exp(psAi+psGBi);

bb=G_As .* H_Bs .* exp(psGAi) - ...
   G_Bs .* H_As .* exp(psGBi+psHAi-psHBi);

c=C0./ss .* (1 - w .* aa./bb);

return
end
```

Index

a

advection   19, 20
  advection time scale   33
  effect on chemical distribution
      41–43
advective–dispersive–reactive (ADR)
      equation   25–29
  ADR models   30
  boundary conditions   31, 32
  derivation using mass balance
      25–27
  initial conditions   30, 31
  nondimensionalization   32–34
  reaction submodel   27
  solution method using Laplace
      transforms   129–135
  solution method using Fourier and
      Laplace transforms   149–155
  solutions, 1-D   129–135, 137–140,
  solutions, 3-D   149–155, 157–160
  sorption submodel   28, 29
  superposition   38–40
analytical method   3
AnaModelTool   40, 41, 72, 141–143,
      159, 160
  Model 101   175, 176
  Model 102   31–42, 64–66, 176–178
  Model 103   178, 179
  Model 104   46, 47, 125, 179, 180
  Model 104M   180–182
  Model 105   182, 183

  Model 106   44, 45, 48–50, 52,
      54, 55, 57, 59–64, 184, 185
  Model 107   185, 186
  Model 108   58, 187–189
  Model 109   58, 189–191
  Model 201   191–193
  Model 202   193–195
  Model 203   195–197
  Model 204   197–199
  Model 205   200, 201
  Model 206   201–203
  Model 207   203–205
  Model 208   206, 207
  Model 301   73–76, 207–209
  Model 302   210–212
  Model 303   212–215
  Model 304   215–217
  Model 305   217–220
  Model 306   220, 221
  Model 401   222, 223
  Model 402   223–225
  Model 403   121, 225–229
  Model 405   229–231
  Model 406   232–234
  user instructions   141–143
aquifer   2
Aris' method of moments   88, 97,
      163, 164
Aris' modified method of moments
      103, 165, 167–169
artesian aquifer   10

Analytical Modeling of Solute Transport in Groundwater: Using Models to Understand the Effect of Natural Processes on Contaminant Fate and Transport, First Edition. Mark Goltz and Junqi Huang.
© 2017 John Wiley & Sons, Inc. Published 2017 by John Wiley & Sons, Inc.
Companion Website: www.wiley.com/go/Goltz/solute_transport_in_groundwater

b

Borden field experiment   124–126
  breakthrough curve behavior   125
  parameters used to model   125
  spatial and temporal sampling   124,
    125
  spatial moment behavior   125, 126
boundary conditions
  Cauchy/flux boundary condition
    32
  Dirichlet/concentration boundary
    condition   31
  effect on breakthrough curves
    64–66
  Neumann boundary condition   32
  volume-averaged resident
    concentration *vs.* flux-averaged
    concentration   66, 67
Boussinesq equation   13

c

Cauchy/flux boundary condition   32
characteristic scale
  concentration   xi, 33
  length   xi, xii, 33, 58
  time   33, 98
chemisorption   22
conceptual model   1
confined aquifer   10
contaminant remediation   121–123
  cleanup time sensitivity analysis
    121–124
  effect of sorption rate on cleanup
    time   122–124
  Laplace domain solutions   121,
    171–174

d

Damköhler number   xi, 33, 34, 54, 56,
    60–62, 81, 83–85, 89–92, 94, 95,
    96, 99–101, 112, 115
Darcy's law   3, 4, 8, 9
Darcy velocity   4

degradation kinetics   27
  Damköhler number for degradation
    xi, 33, 34, 60–62, 83–85, 89–92,
    94, 96, 99, 101
  effect on spatial moments   104–120
  effect on temporal moments
    88–102
  first-order degradation   24, 25,
    60–64, 83–85
  solution to ADR with first-order
    degradation   149–151
  solution to ADR with zeroth-order
    degradation   153–155
  timescale   33
  zeroth-order degradation   24, 25
Dirac delta function   xii, 31, 45, 48, 71,
    76, 81, 82, 89, 104, 105
Dirichlet/concentration boundary
    condition   31
dispersion   20–22, 43–48, 73–78
  effect on breakthrough curves vs
    spatial distributions (1-D)
    43–48
  effect in three dimensions   73–78
  Gaussian (normal) concentration
    profiles   46–48
  molecular diffusion and mechanical
    dispersion   21
  Peclet number   xii, 33, 44, 47, 64,
    67, 73, 76, 91, 93–95, 102
  scale effect   22
  timescale   33, 34, 44, 73
dispersivity   21
  relation to cleanup time   122
drawdown   14, 15

e

equilibrium sorption (also see sorption
    distribution coefficient)   22, 28
  breakthrough curve behavior
    48–49, 78, 79
  linear vs nonlinear   22

related to desorption rate constant
23, 48

spatial distribution behavior   50,
79, 80

spatial moment behavior   104–106

temporal moment behavior   88–97

f

fate and transport processes

advection   19, 20

chemical and biochemical processes
24, 25

dispersion   20–22

sorption   22–24

Fick's first law of diffusion   21

Fick's second law of diffusion   21, 23

flow equation   5–7

derivation using mass balance   6, 7

steady-state   7

flux-averaged concentration   66, 67

relation to resident concentration
67

Fourier transforms   103

application to solve
three-dimensional ADR
equation   71

first-order degradation kinetics
149–151

zeroth-order degradation kinetics
153–155

g

Gaussian concentration profiles
46–48

h

half-life   25

Heaviside step function   xi, 38, 39

Homogeneous/heterogeneous medium
4

hydraulic conductivity   xi, 4

hydraulic gradient   xi, 3, 4

hydraulic head   xi, 3, 4

i

isotropic/anisotropic medium   4

l

Laplace transforms   145–148

definition   129

inverse transform   40, 130, 131, 134,
135, 151, 154, 165, 167

used in AnaModelTool software
40, 41

used to evaluate temporal moments
88

used to solve one-dimensional ADR
equation

first-order degradation kinetics
129–132

zeroth-order degradation kinetics
133–135

used to solve radial ADR equation
171–173

used to solve three-dimensional
ADR equation

first-order degradation kinetics
149–151

zeroth-order degradation kinetics
153–155

linear equation   13

m

mass flux   21

model

analog model   1

conceptual model   1

definition   1

importance of assumptions   12, 22,
25, 82

mathematical model   1

model code   3

model solution   3

physical model   1

purpose   1

moments

spatial

moments (*contd.*)
  absolute   102
  behavior
    first spatial moment behavior
      106, 107, 112–115
    second spatial moment behavior
      107–110, 115–119
    zeroth spatial moment behavior
      106, 110–112
  central   103
  definition
  absolute   102
  central   103
  normalized   103
  evaluation using modified Aris'
    method of moments   103, 165,
    167–169
  temporal
  behavior
    first temporal moment behavior
      91, 92, 98
    second temporal moment
      behavior   92–97
    zeroth temporal moment behavior
      89–91, 98–102
  definition
    absolute   87
    central   88
    normalized   87
  evaluation using Aris' method of
    moments   88, 97, 163, 164

n
Neumann boundary condition   32
nondimensionalization   32, 34
nonlinear models   13, 24
numerical methods   3

p
Peclet number   xii, 33, 44, 47, 64, 67,
  73, 76, 91, 93–95, 102
physical model   1

pore velocity   xii, 19, 20
porosity   5

r
rate-limited sorption   22–24, 28, 29,
  51–60, 80–83, 97–102,
  105–119
  breakthrough curve behavior
    51–53, 81–83, 125
  Damköhler number for sorption   xi,
    54, 56, 81, 99, 100, 112, 115
  diffusion-limited   23, 24, 29, 57–60,
    equivalent first-order rate
      constants   57–60
    mobile–immobile model   23, 24,
      29–31, 58–60
  first-order kinetics   22–23, 28–29,
    51–57, 80–83, 97–102, 105–119
  relation to cleanup time   125, 126
  spatial distribution behavior
    53–57, 80–81
  spatial moment behavior   106–119,
    126
  temporal moment behavior
    97–102
  timescale   22, 54, 56, 99
resident concentration   66, 67
retardation factor   xii, 28, 48, 53, 58,
  63, 78, 92, 93, 96, 97, 99, 104,
  106, 107, 122
  ratio of total mass to dissolved mass
    50, 51, 56
  relation to cleanup time   121–123
Reynold's number   4

s
semianalytical method   3
sorption
  adsorption   22
  chemisorption   22
  sorption equilibrium (see
    equilibrium sorption)
  sorption kinetics (see rate-limited
    sorption)

sorption distribution coefficient xii, 22, 23, 44, 51, 52, 58, 125
   related to desorption rate constant 23, 48, 51, 52
   related to retardation factor xii, 28, 48, 51
specific discharge 4
specific storage 7
superposition 13–15, 38–40

t

tailing
   breakthrough curve 52, 53, 59, 60, 82, 83, 98, 100, 123–125
   spatial distribution 53, 54, 81
Thiem equation 12

u

unconfined aquifer 10

v

virtual experimental system 41, 72, 73

w

water content 5, 6
water table aquifer 10

z

zeroth-order degradation kinetics 37, 38, 71, 133–135, 153, 154